Technology and Writing

Readings in the Psychology of Written Communication

Books of related interest

Education and Informatics Worldwide:
The State of the Art and Beyond
Edited by Frank Lovis
ISBN 1 85302 089 3

Learning Difficulties and Computers:
Access to the Curriculum
David Hawkridge and Tom Vincent
ISBN 1 85302 131 8

Information Technology Training for People with Disabilities
Edited by Michael Floyd
ISBN 1 85302 129 6

Technology and Writing

Readings in the Psychology of Written Communication

Edited by James Hartley

Jessica Kingsley Publishers
London and Philadelphia

First published in the United Kingdom in 1992 by
Jessica Kingsley Publishers Ltd
118 Pentonville Road
London N1 9JN

British Library Cataloguing in Publication Data
Technology and writing : readings in the psychology of written communication.
1. Written communication. Psychology
I. Hartley, James *1940-*
153.6

ISBN 1-85302-097-4

Printed and bound in Great Britain by
Biddles Ltd, Guildford and King's Lynn

Contents

Part IV. Writing for electronic text

Part V. Computer-aided writing: future trends

Preface

Ten years after publishing *The Psychology of Written Communication* I decided, in August 1991, that it would be an interesting venture to produce a second edition. After all, much had happened in the intervening period. Indeed, 1980 seemed to be the year that signalled the start of a flood of books on writing, a flood which has continued unabated. Now, I wondered, where had we got to, ten years later?

I began by reviewing research in much the same categories that I had used in my earlier text, but I found that this did not really work. Each interesting advance seemed to involve a discussion of computers and writing. (Computers had hardly figured in my 1980 text - there being only two papers on the topic.) So I decided to change the focus of the text and to look at two issues: how did technology affect (1) the process of writing, and (2) the resulting product?

To answer these questions this text is divided into five parts. Part I is introductory. Part II looks at computers and writing in an educational context. Part III examines computers and writing in what I have termed special circumstances. Part IV looks at difficulties in writing *for* computers, as opposed to writing with them. And, finally, Part V looks to the future.

One underlying question (not resolved) throughout the text is whether or not technological advances will change writers' capabilities or, more likely, whether or not they will change the nature of the tasks that are performed by writers. Some people assume that word-processors, for instance, will make no difference to the quality of writing, but will change the way in which the task is approached. My own optimistic view is that both might happen: the methods by which written communication is achieved will change, and the resultant quality will improve. Whatever in fact does happen, it is clear that the nature of writing and the way that it is taught will change profoundly in the next decade.

I am grateful to many people for their assistance in the preparation of this text, not least to my fellow contributors, as well as to their editors and publishers. Ahmad Beh-Pajooh deserves my especial thanks for encouraging me to create this sequel to *The Psychology of Written Communication*, and so too do Margaret Woodward, Doreen Waters and Jenny Everill for word-processing my own special brand of written communication.

James Hartley
August, 1991

Part I

Introduction

Chapter 1

Introduction

James Hartley

Part I

Technology & Writing is divided into five main parts. In this first part there is a brief overview of the contents of each part, and an introductory review of previous research on writing. This review, which aims to provide a backcloth for the remaining chapters in this book, focuses on three issues: the nature of writing; learning to write; and evaluating the quality of the written product.

Part II

Part II presents chapters on educational aspects of technology and writing. The examples given here are from schools and colleges and focus on the learners. (British readers might care to note that if they add five to the grade-levels reported by these American authors they will arrive at the age-level of the learners involved.)

One of the key issues raised in Part II is how, perhaps surprisingly, the introduction of technology into the writing process has increased the study of how writers improve their writing skills through collaboration. In Chapter 3 Colette Daiute shows—via the method of a case-history—just how difficult it is in fact to assess all that is going on when pupils write in pairs. However, collaboration in writing is becoming widespread, especially when one is writing on a grand scale for computer presentation (see for example Barrett, 1989a; Irish and Trigg, 1989; Rada et al, 1989).

In Chapter 4 Daiute and Kruidenier take a different tack. This chapter shows how computers can provide prompts to help the solitary writer at his or her task. The computer can in effect act as a collaborator. The question raised is whether or not the computerised prompts can become internalised by the writers and form part of their own self-regulated writing skills. Other investigators have examined the use of prompts in this way for teaching writing, but not always in a computer-based context (see e.g. Bereiter & Scardamalia, 1987; Beal, Garrod & Bonitatibus, 1990; Brown & Day, 1983; Graham & Harris, 1989; Wallace & Bott, 1989).

In Chapter 5 Susan Zvacek provides a general review of the effects of word-processors on writing skills. Zvacek examines how word-processing (1) influences the composing process and (2) affects the quality of the final product. In both areas the evidence for particular effects is difficult to obtain, and there is much debate. These issues are discussed further in this chapter. In addition to these concerns Zvacek comments on two more important points: these are (1) the effects of motivation, and how writing with computers seems to enhance this for most (but not all) writers; and (2) the problems of equality of opportunity. It is clear that progress is restricted in both the academic and the business world if one does not have access to the appropriate technology. This problem, of course, is also germane to Part III of this book where we consider how technological advances can aid handicapped people to express their views in writing.

In the final chapter in this part (Chapter 6) Stephen Bernhardt and his colleagues present a detailed account of one particular study of teaching composition with computers. A great many measures were made in this study over a one term period. The results were generally positive, but Bernhardt and his colleagues introduce two more important considerations into the discussion: (1) they stress the importance of the teacher's role (and this is analysed in detail in a later paper—Bernhardt et al, 1990); and (2) they indicate that students don't necessarily like interacting together when they are learning to write. Other papers, too, sometimes confound one's expectations. Peacock and Breese (1990), for example, found that their eleven-year-olds preferred to write by hand rather than learn to use a lap-top word-processor.

The educational context for work on computers and writing is very broad. Regretfully, limitations of space have prevented me from including chapters that describe research on academic or occupational writing by adults. However, the papers by Doheny-Farina (1988a), Krull (1989), Teles and Ragsdale (1989) and the books by Anderson, Brockmann and Miller (1983) and Doheny-Farina (1988b) may be of interest here. Also, I have suggested additional reading at the end of most of the chapters in this book so that readers can follow up particular points of detail, should they wish to do so.

Part III

The issues raised above also appear in Part III, with slightly different emphases. Collaboration, motivation, the importance of the teacher, and the use of computer prompting are all important issues when we consider how we can use technology to help those less able than ourselves. And here, of course, we can consider technology in a wider sense. The adaptations of tape-recording systems, video and computer technology for the blind, the deaf and the physically handicapped provide obvious examples. But there are also more specialised tools, such as manual and electronic braille machines, moon writers, large print for the visually handicapped, and synthesised speech, audio amplification devices, and techniques for putting captions on TV for the hearing handicapped (See e.g. Alpiner & Vaughan, 1988; Edwards, 1989; Hartley, 1989; Maley, 1987; Mercer et al., 1985; Smith et al., 1989).

In Chapter 7 Charles MacArthur outlines some of the difficulties that learning disabled children experience in learning to use word processors, and some of the advantages (especially for those with poor handwriting skills). Motivation and social support again feature as important issues, as do interactive prompting programs and teaching support (e.g. see Graham and MacArthur, 1988; MacArthur et al., 1990).

In Chapter 8, Keith Nelson, Philip Prinz and Deborah Dalke present a recent account of what is known as the ALPHA project. This project is concerned with using an interactive program for Apple II computers to teach reading and writing skills to deaf children. As such it is one of the few projects available that examines how computers might be used to help develop the writing skills of deaf or partial hearing students. Key components of the program are the use of a touch keyboard that displays words rather than letters, and the use of sign language on screen and by the teachers.

Chapters 9 and 10 examine the use of technology with blind and visually handicapped students in the context of learning to write. Richard Ely in Chapter 10 gives an overview of the issues, and Alan Koenig and his colleagues describe in Chapter 11 the start of a project using computers to teach writing with visually handicapped children. Regrettably the authors of this report have been unable to supply me with any more details concerning the project and its achievements. It is likely that much more work is going on with new techniques for handicapped students than is actually written about. Much of this work is tailored to individuals and carried out single-handedly by enthusiastic teachers. The difficulties of doing systematic research in these areas are formidable and this may account for why so little has been published on it in academic journals.

I am not sure that senior citizens would like to be bracketed with the handicapped—and I apologise to any such person who might be reading this text. However, Sydney Butler's contribution (Chapter 11) is important because it not only focuses on the writing of the elderly, but it also shows how computer technology might be used to enhance it. Here, as for the blind and the deaf, there is little literature on technology and writing. There is actually quite a literature on writing *for* the elderly (see the suggested further reading after Chapter 11) but there is not so much on the writing *of* the elderly. Butler's chapter is also of interest because it again stresses how much can be achieved by making writing a social process—as do other books and papers on the writing of the elderly (e.g. Butler, 1988; Gotterer, 1989; Supiano et al, 1989).

Butler reports (personal communication) that the lifewriting group described in Chapter 11 continues to meet weekly for writing and discussion. In the five years since the group began the members have established themselves as an autonomous organisation, and they have published four more anthologies of their stories and poems using microcomputers and desk-top publishing. Moreover, several members of the group have self-published their own collections as family histories.

Part IV

So far we have considered using technology to aid the writing process. In Part IV, however, we turn to consider problems of writing *for* the computer. A great deal of the text that we read today (and even more that we shall read tomorrow) is presented on a computer or television screen. The process of writing for such screen-based presentations is markedly different from that used for conventionally printed text, mainly because the current screens are much smaller than most printed pages. Writing for small screens implies that one has to think in smaller chunks. One has to consider how sequencing, the use of headings, and other text features—such as tables and illustrations—will present particular problems. (For further elaborations on these points, see Wright, 1987a,b, 1989.) Once, when I was writing an article for an electronic journal (Pullinger, 1984),

I had to make a decision about a table of results. Should I simplify it to form one 'screenful', divide it up into three 'screenfuls', place it in an Appendix where readers could refer to it if they wished, or should I omit it altogether? I chose in the end to simplify it.

In Chapter 12 Gary Morrison and his colleagues consider these issues of *text density* (the number of concepts per 'screenful') and *screen density* (the number of character spaces used) on today's conventional screens. In the light of my comments above about simplification I was particularly interested to observe in Morrison's chapter that their students opted for well-filled rather than spacious screens—contrary to what one might predict from the literature on screen design (see Hartley, 1987; Tullis, 1988).

Today, of course, there are larger screens available than the conventional ones used by Morrison and his colleagues. It is possible to display two A4 pages side by side and still to have space for innumerable 'windows'. (In passing we may note with Edwards (1989) that windows, icons, menus and pointer systems will provide significant new obstacles for the visually handicapped.) If screen sizes do change, of course, then much of the work of the kind done by Morrison et al will need to be repeated. Furthermore, we may well expect more research on the design of windows, icons and menus (see, e.g. Billingsley, 1988; Paap & Roske-Hofstrand, 1988) and on the electronic presentation of graphic and tabular materials (see e.g. Lefrere, 1989; Norrish, 1984).

One way to help writers organise their text for screen or desk-top publishing is to provide 'templates' for them to follow. Templates help writers to keep control over their text and to use consistent formats. Readers (and writers) benefit from such consistency in layout: readers do not want to think every few moments 'now where do I go from here?'. There are similar discussions of such template procedures for automating the layout of tables (e.g. see Lefrere, 1989) and the presentation of forms (e.g. see Frohlich, 1987). David Kember in Chapter 13 describes the uses of templates in the development of printed study materials. Devices such as these that help readers and writers to navigate text are important, especially when it comes to writing and presenting hypertexts (Chapters 14 and 15).

The term 'hypertext' is used to describe a database of related text fragments that the user can choose to read in any order. Jonassen (1989) introduces his book on hypertext as follows:

> This book is a hypertext about hypertext and hypermedia. Hypertext is among so many other things a form of non-sequential writing. Hypertext is also a collection of text fragments rather than continuous prose. These text fragments are tied together by links. It is not meant, like most texts, to be read from beginning to end. You may read it that way, but to use it as a hypertext you will make choices about what you read next.

Figure 1.1 shows what is in effect the contents page of Jonassen's book. If you turn, say, to page 5 to study the characteristics of hypertext, you will find another circular 'hypermap'. Here you can choose to read any one of fourteen more text fragments in any order that you decide. And, when reading these fragments, you can cross refer to other 'pages'. The main topics, if you like, are nodes (or points of departure for several different issues) and each node and issue has connecting links across the text.

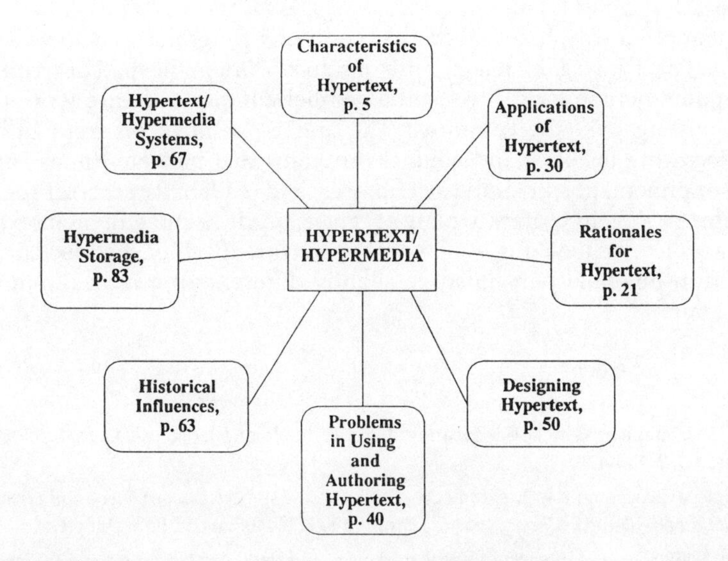

Figure 1. The 'contents' page from Hypertext/Hypermedia. *Figure reproduced with permission of the author and the copyright holder, Educational Technology publications.*

Most exemplars of hypertext are computer-based and information is presented on an electronic screen. Computer-based hypertexts can contain textual, graphic and audiovisual data stored on video and compact discs. Hypertexts can range from simple systems, such as Jonassen's, to vastly complex ones, such as the procedures manual for the American nuclear-powered aircraft carrier, the USS *Carl Vinson*. Some suggested further reading on this topic is provided at the end of Chapter 15.

Hypertext formats seem particularly suitable for dictionaries and encyclopaedias, instructional manuals, research handbooks, and conference proceedings. In essence, as Zimmerman (1989) puts it, 'in hypertext applications, the metaphor of the network has replaced that of the hierarchy. A network has no top or bottom, but a plurality of connections that increase possible interactions between components of the network. Hypertext writing is non-sequential... and the skills required to design such information products are "a giant leap" beyond those we currently use.'

Ben Shneiderman describes in Chapter 14 some of these problems of writing for hypertexts (This chapter is in fact an excerpt from a lengthier chapter in the text edited by Edward Barrett in 1989, a text that well repays reading in this context). A more detailed account of a specific application of a hypertext system (creating an electronic encyclopaedia) is provided by Kreitzberg and Shneiderman in Chapter 15. References to other applications are also provided in the list of further reading following this chapter.

Part V

Part V returns to more conventional issues—in the sense that none of the authors in this part refer to hypertext or hypermedia. The issue of debate in Part V is how one can use computer programs to aid the writing process, or perhaps, putting it another way, how can one use computer programs to help writers to think. If in Part IV we gained a glimpse of future work in the field of writing *for* the computer, here in Part V we gain a parallel glimpse of future work in the field of writing *with* the computer. The aim is to move on from utilising word-processing facilities and adjunct programs that indicate things such as spelling or punctuation errors, sexist phrases, and readability scores. Programs of the future will help writers to plan, organise, produce and edit their text. Just how this will be achieved is a matter of conjecture. Each of the four chapters contained in this final part illustrate slightly different approaches to this important issue.

References

Alpiner, J. G. & Vaughan, G. R. (1988). Hearing, aging, technology. *International Journal of Technology and Aging*, 1, 2, 126–135.

Anderson, P. V., Brockmann, R. J. & Miller, C. R. (Eds.) (1983). *New Essays in Technical & Scientific Communication: Research, Theory, Practice*. Farmingdale, NY: Baywood Publishing Co.

Barrett, E. (1989a). Textual intervention, collaboration, and the online environment. In Barrett, E. (Ed.), *op. cit.*

Barrett, E. (Ed.) (1989b). *The Society of Text*. Cambridge, Mass.: MIT Press.

Beal, C. R., Garrod, A. C. & Bonitatibus, G. J. (1990). Fostering children's revision skills through training in comprehension monitoring. *Journal of Educational Psychology*, 82, 2, 275–280.

Bereiter, C. & Scardamalia, M. (1987). *The Psychology of Written Composition*. Hillsdale, NJ: Erlbaum.

Bernhardt, S. A., Wojahn, P. G. & Edwards, P. R. (1990). Teaching college composition with computers: a timed observation study. *Written Communication*, 7, 3, 342–374.

Billingsley, P. A. (1988). Taking panes: issues in the design of windowing systems. In Helander, M. (Ed.) *Handbook of Human-Computer Interaction*. Amsterdam: Elsevier.

Brown, A. & Day, J. D. (1983). Macro-rules for summarising texts: the development of expertise. *Journal of Verbal Learning and Verbal Behavior*, 22, 1–14.

Butler, S. (1988). *Lifewriting: Self-Exploration and Life Review through Writing*. Dubque, Iowa: Kendall/Hunt.

Doheny-Farina, S. (1988a). Writing in an emerging organisation: an ethnographic study. *Written Communication*, 3, 2, 158–185.

Doheny-Farina, S. (1988b). *Effective Documentation: What Have We Learned From Research?* Cambridge, Mass.: MIT Press.

Edwards, A. D. (1989). Soundtrack: an auditory interface for blind users. *Human Computer Interaction*, 4, 1, 45–66.

Frohlich, D. M. (1987). On the re-organisation of form filling behaviour in an electronic medium. *Information Design Journal*, 5, 111–128.

Gotterer, G. M. (1989). Storytelling: a valuable supplement to poetry writing with the elderly. *Arts in Psychotherapy*, 16, 2, 127–131.

Graham, S. & Harris, K. R. (1989). Improving learning disabled students' skills at composing essays: self-instructional strategy training. *Exceptional Children*, 56, 3, 201–214.

Graham, S. & MacArthur, C. (1988). Improving learning disabled students' skills at revising essays produced on a word processor: self instructional strategy training. *Journal of Special Education*, 22, 133–152.

Hartley, J. (1987). Designing electronic text: the role of print-based research. *Educational Communication and Technology Journal*, 35, 1, 3–17.

Hartley, J. (1989). Text design and the setting of braille (with a footnote on Moon). *Information Design Journal*, 5, 3, 183–190.

Irish, P. M. & Trigg, R. H. (1989). Supporting collaboration in hypermedia: issues and experiences. In Barrett, E. (Ed.), *op. cit.*

Jonassen, D. H. (1989). *Hypertext/Hypermedia*. Englewood Cliffs, NJ: Educational Technology Publications.

Krull, R. (1989). Online writing from an organisational perspective. In Barrett, E. (Ed.), *op. cit.*

Lefrere, P. (1989). Design aids for constructing and editing tables. *British Library Research Paper No. 61*. London: British Library

MacArthur, C. A., Haynes, J. A., Malouf, D. B., Harris, K., & Owings M. (1990). Computer assisted instruction with learning disabled students: achievement, enjoyment and other factors that influence achievement. *Journal of Educational Computing Research*, 6, 311–328

Maley, T. (1987). Moon à la mode. *New Beacon*, LXXI, 840, 109–113.

Mercer, D., Correa, V. L. & Jowell, V. (1985). Teaching visually impaired students word processing competencies: the use of the Viewscan Textline. *Education of the Visually Handicapped*, 17, 1, 17–29.

Norrish, P. (1984). Moving tables from paper to screen. *Visible Language*, 18, 2, 154–170.

Paap, K. R. & Roske-Hofstrand, R. S. (1988). Design of menus. In Helander, M. (Ed.) *Handbook of Human-Computer Interaction*. Amsterdam: Elsevier.

Peacock, M. & Breese, C. (1990). Pupils with portable writing machines. *Educational Review*, 42, 1, 41–56.

Pullinger, D. J. (1984). Design and presentation of Computer Human Factors journal on the BLEND system. *Visible Language*, 23, 171–185.

Rada, R., Keith, B., Burgoigne, M. & Reid, I. (1989). Collaborative writing of text and hypertext. *Hypermedia*, 1, 2, 93–110.

Smith, A. K., Thurston, S., Light, J., Parnes, P. et al. (1989). The form and use of written communication produced by physically disabled individuals using microcomputers. *AAC—Augmentative and Alternative Communication*, 5, 2, 115–124.

Supiano, K. P., Ozminkowski, R. J., Campbell, R. & Lapidus, C. (1989). Effectiveness of writing groups in nursing homes. *Journal of Applied Gerontology*, 8, 3, 382–400.

Teles, L. & Ragsdale, R. (1989). The impact of word processing on faculty writing behaviour. *Higher Education Research and Development*, 8, 2, 217–235.

Tullis, T. S. (1988). Screen design. In Helander, M. (Ed.) *Handbook of Human-Computer Interaction*. Amsterdam: Elsevier.

Wallace, G. W. & Bott, D. A. (1989). Statement pie: a strategy to improve the paragraph writing skills of adolescents with learning disabilities. *Journal of Learning Disabilities*, 22, 9, 541–543, 553.

Wright, P. (1987a). Writing technical information. In Rothkopf, E. Z. (Ed.), *Review of Research in Education. Vol. 14*. Washington: American Educational Research Association.

Wright, P. (1987b). Reading and writing for electronic journals. In Britton, B. K. & Glynn, S. M. (Eds.), *Executive Control Processes in Reading*. Hillsdale, NJ: Erlbaum.

Wright, P. (1989). Interface alternatives for hypertexts. *Hypermedia*, 1, 2, 146–166.

Zimmerman, M. (1989). Reconstruction of a profession: new roles for writers in the computer industry. In Barrett, E. (Ed.), *op. cit.*.

Chapter 2

Writing: A Review of the Research*

James Hartley

My aim in this chapter is to provide a brief review of some of the main issues covered in research on writing in order to provide a background for the chapters that follow in this text. Readers who are interested in more detailed reviews are referred to the papers by Applebee, 1984; Durst & Newell, 1989; Fitzgerald, 1987; Freedman et al., 1987; Hayes & Flower, 1986; Humes, 1983; and Huot, 1990. There are, of course, many books and book chapters on the topic, and several of these will be referred to in this review.

For convenience of presentation the review is divided into three overlapping sections. Two of these sections, *the nature of writing* and *learning to write*, are discussed in detail by Freedman et al. (1987). It seemed appropriate to add a third concern, *evaluating the quality of writing*, in light of the issues raised in this text. A fourth topic, *the uses of writing*, is covered in more detail by Freedman et al.; readers who are interested in the social aspects of writing and how cultures and activities are shaped and have been shaped by writing, are referred to this paper.

The nature of writing

Developments in computer technology, hand in hand with developments in cognitive psychology, have had a profound effect on research on writing. The current research addresses different issues and different questions from those asked some fifty, or even twenty years ago. Nonetheless, this does not mean that all of this early research is irrelevant: there is a past literature which it is fruitful to explore in considering where we are now, and where we are going next.

If one examines early educational psychology textbooks and those of the present day it is clear that there has been a shift of emphasis in educational practice as far as writing is concerned. In the 1920s the emphasis was on handwriting skills, in the 1950s it was on the grammatical quality of the written products, and in the 1990s it is on the *process* of writing—how writers arrive at their end products.

Indeed, it is now common-place to talk of 'process-approaches' to teaching writing. Thus, instead of analysing classic examples of good practice, and learning the rules that govern such practice, process-approaches focus on allowing the learners to explore sub-components of the writing process in a variety of different contexts.

---

* *This chapter has been specially prepared for this book.*

Bizzell (1986) describes how research on the process-approach to teaching writing began in the USA with the work of Emig (1971) and in Britain with the work of Britton et al. (1975). Emig noted at the time, 'of the 504 studies written before 1963 that are cited in the bibliography of *Research in Written Composition* only two deal even indirectly with the process of writing among adolescents.' So her work marked a significant change of emphasis.

Britton et al. and Emig worked independently, but they were aware of each other's work. One of the key ideas that they shared between them was that the processes used to produce text depended upon the nature of the text. Britton et al. distinguished between three text genres along a continuum: poetic (written for oneself), expressive (written for an intimate friend) and transactional/expository (written for a less personal audience, such as the teacher). Emig described two kinds of writing—'reflexive' (paralleling Britton's expressive) and 'extensive' (paralleling Britton's transactional).

In carrying out their research, Britton et al. examined about 2000 essays written by British schoolchildren aged between eleven and eighteen, and Emig interviewed eight nineteen-year-olds about their writing and provided a detailed case-history, including think-aloud protocols, of the writing processes of one of them. Britton et al. and Emig found that their participants generally spent far more time on transactional writing than on expressive writing, but Emig found that her American students much preferred, and spent more time (at home) on expressive writing. School (transactional) writing was well-learned but it was a mechanical activity. Bizzell (1986) concludes that these influential research workers brought to the fore two important themes: (1) the process of writing depends upon the nature of the text; and (2) to study writing one needs to examine writers at work.

Table 2.1 below suggests that there are indeed numerous kinds of writing. Writing has different purposes, audiences and genres. And, as we shall see in the following sections of this review, this is important when it comes to examining the acquisition and evaluation of writing skills.

There have been many studies of writing processes following this pioneering work (see Freedman et al. for references). However, the most influential approach, and the one most frequently referred to in the chapters in this textbook, is that of Hayes and Flower (see, e.g. Hayes & Flower, 1980a,b; 1983; 1986).

Hayes and Flower used think-aloud protocols to develop their cognitive process model of writing. They described the method as follows

> In protocol analysis the subject is asked to perform a task and to 'think aloud' while performing it. In think-aloud studies of writing the writer is asked not only to think aloud but to read and write aloud as well . . . The data from think-aloud studies are contained in the verbatim transcript of the tape-recording (with all the um's, pauses and expletives), together with the essay and all the notes that the writer has generated along the way. The transcript is called a protocol. These materials are then examined in considerable detail for evidence that may reveal something of the processes by which the writer has created the essay. In general the data are very rich in such evidence. Subjects typically give many indications of their plans and goals, e.g. 'I'll just jot down ideas as they come to me'; about strategies for dealing with the audience, e.g., 'I'll write this as if I were one of them', or about criteria for

Table 2.1

Dominant Intention/ Purpose	Primary Audience	Cognitive Processing: Reproduce		Organize/Reorganize		Invent/Generate	
	Primary Content	Facts	Ideas	Events	Things, facts, mental states, ideas	Ideas, mental states, alternative worlds	
To learn (metalingual)	Self	Copying, taking dictation		Retell a story (heard or read)	Note Resume Summary Outline Paraphrasing	Comments on book margins Metaphors Analogies	
To convey emotions, feelings (emotive)	Self Others	Stream of consciousness		Personal story Personal diary Personal letter	Portrayal	Reflective writing – Personal essays	The traditional literary genres and modes can be placed under one or more of these four purposes
To inform (referential)	Others	Quote	Fill in a form	Narrative report News Instruction Telegram Announcement Circular	Directions Definition Technical description Biography Science report/ experiment	Expository writing – Definition – Academic essay/article – Book review – Commentary	
To convince/ persuade (connative)	Others	Citation from authority/expert		Letter of application Statement of personal views, opinions	Advertisement Letter of advice	Argumentative/ persuasive writing – Editorial – Critical essay/article	
To entertain, delight, please (poetic)	Others	Quote poetry and prose		Given an ending– create a story Create an ending Retell a story	Word portrait or sketch Causerie	Entertainment writing – Parody – Rhymes	
		Documentative discourse		Constative discourse: Narrative; Explanatory	Descriptive	Exploratory discourse: Interpretive (Expository/ Argumentative/ Persuasive)	Literary

evaluation, e.g. 'We'd better keep this simple'. The analysis of this data is called protocol analysis (Hayes & Flower, 1986).

Using protocol analysis, Hayes and Flower constructed the process model of writing shown in Figure 2.1. It can be seen that there are two major areas that impact on the central 'box': the task environment, and the writer's long term memory.

The *task environment* includes, in addition to the setting, the writing assignment—the topic and the intended audience—and structural constraints (such as the amount of time available). The task environment also includes the text which the writer has produced so far. This is an important part of the environment because the writer refers to it repeatedly whilst composing.

The writer's *long term memory* allows the writer to produce the content based on his/her own knowledge (without constantly having to look things up) and it includes, for the expert writer, a familiarity with spelling, grammar and writing genres—such as those outlined in Table 2.1.

The task environment and the writer's long term memory provide the context in which the writing processes operate. Here, there are three key elements: planning, translating and reviewing.

Planning involves generating content, organising it and setting up goals and procedures for writing. Hayes & Flower (1983) write: 'We see Planning as a very broad activity that includes deciding on one's meaning, deciding what part of that meaning to convey to the audience, and choosing rhetorical strategies. In short, it includes the whole range of thinking activities that are required before we can put words on paper. It is important to note that (1) planning goes on throughout composing and (2) the plan may not be encoded in a fully articulated or even in a verbal form (Plans may be visual images.)'

Translating is the term that Hayes & Flower use to describe the physical act of expressing the content of the planning process in written English. Hayes and Flower (1983) write: 'Translating is the act of expressing the content of Planning in written English. Although one can reliably distinguish when writers move from Planning (which may produce notes and doodles) to Translating (the attempt to produce prose), this does not mean that writers have a fully formed meaning that they simply express in words. Rather, writers have some more or less developed representation encoded in one form. The act of translating this encoded representation to another form (ie written English) can add enormous new constraints and often forces the writer to develop, clarify, and often revise that meaning. For that reason, the act of translating often sends writers back to planning. Often these processes alternate with each other from one minute to the next.'

Reviewing involves evaluating either what has been written or what has been planned. Reviewing often (but not invariably) leads to revision. Hayes and Flower draw a distinction between revision (making substantial changes) and editing (polishing).

Finally, writing must be *monitored*. Hayes and Flower (1983) write: 'The Monitor is the executive of the writing process that determines when to switch from one writing process to another, for example, when one has generated enough ideas and is ready to write. The monitor may function differently from writer to writer and from writing task to writing task. Some people move into

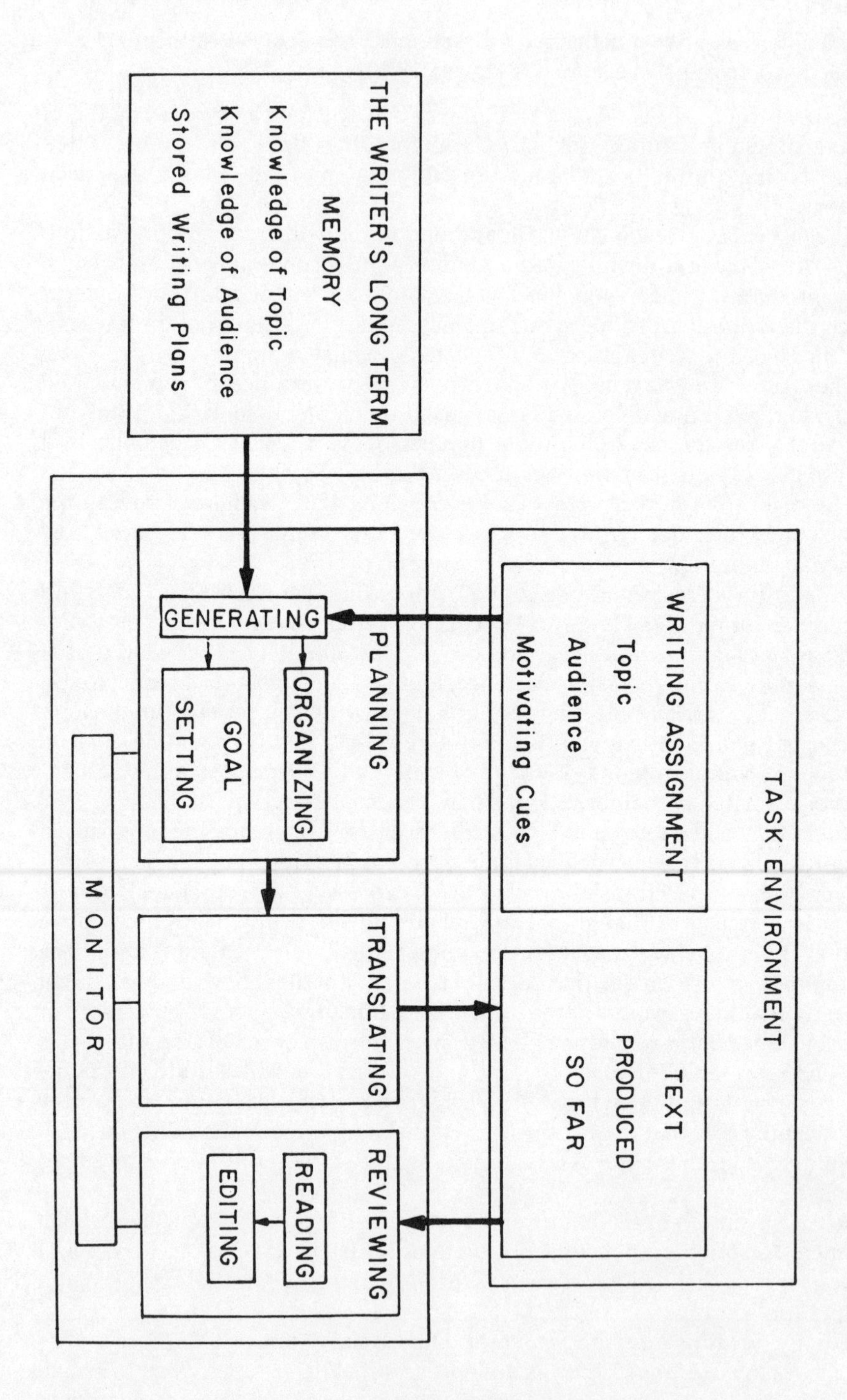

Figure 2.1. The structure of the writing model proposed by Hayes & Flower. Figure reproduced with permission of the authors and the copyright holder, Lawrence Erlbaum Associates.

translation as soon as possible in writing a paper, whereas others will not write a word until they are reasonably certain that planning is complete. Further, a writer undertaking an easy task, for example, a short letter, may do little or no planning before writing the first sentence, whereas the same writer undertaking a difficult task, for example, a philosophical treatise, may plan for months before writing a word.'

Hayes & Flower (1986) provide more detailed descriptions of the processes of planning, translating, and reviewing together with an account of the research that has stemmed from this approach.

Freedman et al. summarise the position as follows:

1. The writing process is a hierarchically organized, goal-directed, problem-solving process.
2. Writing consists of several main processes—planning, transcribing text, reviewing—which do not occur in any fixed order; rather, thought in writing is not linear but jumps from process to process in an organized way which is largely determined by the individual writer's goal.
3. Experts and novices solve the problems posed by the task of writing differently.

As Hayes & Flower (1980b) put it, 'It is no wonder that many find writing difficult. It is the very nature of the beast to impose a large set of converging but potentially contradictory constraints on the writer. Furthermore, to be efficient the writer should attend to all of these constraints at once; when all is said and done they must be integrated. One doesn't, for example, generate all of one's knowledge on a topic and then decide what to do with it. Ideally, each utterance a writer generates would be at once perfectly accurate, well-formed itself, integrated in the text and rhetorically effective. Unfortunately, this ideal rarely occurs because of the limited number of items our short-term memory or conscious attention can handle. Humans are basically serial processors and not well adapted to handling a large number of simultaneous demands on attention. This means that we must handle converging constraints by juggling them in clever ways.'

A good deal of research has been carried out comparing the writing processes of experts and novices. Freedman et al. describe three distinct groups of novices which reviewers do not always distinguish between: (1) students at all levels whose skills are developing; (2) basic writers who are behind their peers or age group; and (3) young writers. Freedman et al. argue that these different groups have different needs.

Freedman et al. indicate that experts create what Hayes and Flower call 'reader-based' prose, whereas novices create 'writer-based' prose. Experts think about how their readers will follow the text, and this helps them to plan their writing better. Novices, on the other hand, tend not to think much about their readers whilst they are writing: they are more concerned instead with the surface features of the text. Very young writers may follow an orderly procedure as they write, but they often lose their train of thought because they have to attend more to mechanical concerns, such as letter formation, handwriting and spelling.

The amount of knowledge that a writer has about a subject also affects the writer's planning. People who are experts in their subject matter will have greater knowledge about it than the non-experts. The problem is, of course, that such subject matter experts may have difficulty in conveying this knowledge to others—especially if they are novice writers. Expert writers, according to Hayes and Flower (1986), create more elaborate plans with more inter-connecting networks than do novices. Schriver (1986) suggests how experts can be taught to predict novice's comprehension problems with text and to adjust their writing accordingly.

A great deal of attention has also been paid to differences between experts and novices in terms of their revision strategies (e.g. see Fitzgerald, 1987; Kurth, 1987). Hayes and Flower (1986) summarise findings of the research on revision as follows:

- there are large differences between writers in the amount of revising that they do. Experts revise more than novices. Expert revisors attend to more global problems (e.g. re-sequencing, re-studying, re-writing large units of text) than do novices
- experts are better than novices at both detecting problems in their text, diagnosing the causes of the problems, and choosing appropriate remedies
- writers find it harder to detect problems in their own texts than they do in other people's texts.

One implication of this last point is that better quality writing might be produced if authors collaborate in revising texts. Another, in line with the argument of this text, is that writers may change their revision strategies if they are assisted by computer technology. Fitzgerald's (1987) review in this respect is not, however, optimistic. She reports that it is not yet obvious whether using a word processor affects either the amount or the quality of the revision carried out.

Be that as it may, such discussions of the nature of writing clearly have implications for teaching and learning. The shift in focus from product to process (outlined at the beginning of this review) has had a profound effect on how writing is taught. The following activities are typically associated with process-approaches to teaching writing: brainstorming; free writing; journal writing; small-group activities; teacher-learner conferences; writing for different audiences; pre-writing; outlining; multiple drafting; deferred revision; the elimination or deferment of assessment; class publication. Teachers who are to the forefront of this approach include Britton (1972), Elbow (1973), Graves (1983, 1984) and Murray (1984a,b.)

Psychological research on the acquisition of cognitive and motor skills would also support the need for separating out different processes, practising, and integrating them. Authors such as Elbow (1973) and Wason (1970; 1980) suggest quick rough drafting, and then final polishing, rather than attempting to produce polished prose at the first attempt, but the evidence to support this is mixed (Glynn et al. 1982, Kellogg, 1988).

However, in concluding this section of this review, we should note that some authors (e.g. Applebee, 1986; Hillocks 1984, 1986) criticize the zealous nature of some advocates of the process-approach to teaching writing. Hillocks (1984), for example, analysed over 500 studies, and compared the effectiveness of three main approaches to teaching writing: (1) student-centred activity-based versions of the process approach; (2) individual instruction, using tutorials or programmed instruction, or a combination of these; and (3) teacher-led structured approaches that often involved enquiry based learning and group problem solving. Hillocks concluded that the third approach was the more effective. Applebee (1986) re-emphasised that Hayes & Flower's descriptive accounts emerged from the analysis of skilled writers solving difficult problems. In school-based situations, Applebee notes, a process-approach to instruction often means providing novice learners with false tasks, which are then practised in isolation from the 'real world'. Thus Applebee considers that, 'rather than suggesting a range of strategies for solving problems, process instruction will become just another series of practice exercises'.

Learning to write

Most children (in the Western world) learn to write at home and in the preschool, using pencils, felt-tipped pens and biros. Few students start with electronic keyboards, although Moore (1962) advocated this in the early 1960s. So, perhaps not surprisingly, the mechanics of handwriting still occupy a great deal of research attention. Clay (1975) published a well-known book in this area, and Bailey (1988) provides a useful review of current research. Bailey covers topics such as the grasp of the pen, pressure on the writing instrument and surface, the use—or not—of lines on writing paper, and the advantages and disadvantages of different writing instruments (such as large lead pencils, triangular pencils, six-sided pencils, biros and felt-tip pens). It appears that current practice is moving away from the use of oversize pencils, that children prefer six-sided pencils if they have to use pencils—but that they would rather use biros than pencils. The emphasis on the mechanics of handwriting is particularly strong in the field of the learning disabled, and Maarse et al. (1991) report on advances in computer recording in this context. Digitizers with pressure-sensitive pens are being developed that allow for the automatic recording and analysis of handwritten scripts.

A concern with the mechanics of handwriting, however, is not the only concern of studies of very young children learning to write. Several researchers have traced the emergence of lettering from children's drawings and scribbles (see e.g. Beard, 1984; Sarachno, 1990). Francis (1990) suggests that parents who encourage scribbling allow their children to develop their manual dexterity as well as an embryonic understanding of authorship. In learning to write their names, and in mimicking adult practices such as writing letters, notes, and shopping lists, children gain practice in the use of upper and lower case letters. Bissex (1980), Chomsky (1971) and Read (1986), amongst others, detail how phonetic forms dominate early written messages. Read (1971) describes how one boy, banished to his room for having hidden his mother's bracelet, sent a series of increasingly conciliatory messages downstairs on paper missiles ending with LOOKC IN THE BAC YRD LOOKC BEHINE THE SHED.

Moving on beyond the preschool, a great many teachers and researchers have discussed developmental changes in the acquisition of writing skills (e.g. see Beard, 1984; Bereiter & Scardamalia, 1987; Farr, 1985; Martlew, 1983). In the United Kingdom there have been a number of observational studies of junior school practice which have commented on the role of writing (e.g. Bennett et al., 1980; DES 1978, 1982; Galton et al., 1980). These studies suggest that primary school children spend about a third of their time on one kind of writing activity or another (although there are wide variations). It appears that much of this writing is of the expressive and narrative kind, rather than expository in nature. Furthermore, much is done for the teacher rather than for any other audience. There are also criticisms that junior school pupils spend too much time copying from textbooks and worksheets, and that they regard neatness as the main criterion for success. In the United States Juel (1988) showed that the probability that a poor reader at age six would remain a poor reader at age nine was very high, but that early writing skills did not predict later writing skills to the same extent.

At the secondary level in the United Kingdom the Assessment and Performance Unit of the Department of Education and Science monitored the writing skills of 11- and 15-year-old pupils (see White, 1986). Many different types of writing were tested, including the pupils' abilities to explain, instruct, narrate, report, give an opinion, describe imaginatively, express feelings satirically, and persuade. The results were assessed by a group of experienced teachers working together with the research team. Each piece of writing was assessed holistically on a scale of 1 to 7, and analytically in terms of content, organisation, appropriateness and style, grammar, and punctuation and spelling. The results showed:

1. Many pupils found writing difficult—although the majority received marks in the middle of the range. Many 15-year-olds were unable to match the best writing achieved by some of the 11-year-olds.
2. There was a wide variation in performance according to the nature of the task. There were few tasks which were uniformly hard or easy for each age group, but the hardest task was the use of an appropriate style. For most 15-year-olds writing a story was easier than making good notes.
3. The pupils' own judgements about how to improve their writing were often based on conceptions of neatness and correctness rather than on matters of subject matter, style, or purpose.
4. Girls generally performed better than boys. More boys than girls of each age group showed negative attitudes towards writing, and negative attitudes were voiced more strongly by the older pupils (see White, 1987).

Parallel results were found in a similar study of 14,000 15-year-olds reported by Gubb, Gorman and Price (1987). In this study, nine different writing types were used, similar to those reported by White (1986). Gubb et al. report that pupils found most difficulty in coping with the demands of expository writing and least difficulty in writing stories. Tasks which were of an intermediate difficulty involved describing a skill or a process, presenting an argument, and explaining how a composition should be written. And, as in White's report, girls did better than boys.

These findings on sex differences are of interest, especially if we think that boys are likely to hold more positive attitudes towards computers than girls. However, such a view may be simplistic. Trueman (1990) reported that, out of nine studies with schoolchildren, three showed boys to be more positive than girls towards computers, five showed no significant differences, and one showed girls to be more positive than boys. However, Trueman also reported six out of ten studies with students showing males to be more positive than females towards computers. It may be that the generally positive attitudes found towards writing with computers (e.g. see Baer, 1988; Shaver, 1990) will enable boys to reduce their negative feelings towards writing.

White (1986) concluded her booklet with a series of recommendations. She suggests that pupils could be helped to improve their writing skills in the following ways:

- by producing less writing in total
- by being encouraged to write for a wider range of purposes and for a more varied readership
- by building on their existing skills as speakers of the language and being made aware of the contrasting demands of speech and writing
- by constructing for themselves, either as individuals or in groups, a definite purpose for what they write
- by being given regular constructive and systematic feedback about the quality of their written work
- by being made aware of the place of writing in the process of learning and of the value of note-making and exploratory drafting; and
- by having opportunities to work collaboratively through stages of composition and revision.

White does not mention computers in her report, but the implementation of some of her recommendations (particularly the last three) could be made more effective through computer-aided composition.

Freedman et al., in their account of the acquisition of writing skills, are at pains to point out the complexity of the process. Their review shows that although it is useful to give simple descriptions and prescriptions of the kind listed by White above, there are deeper concerns. Freedman et al. point to individual variability in pupils' knowledge of writing processes and in the differences between 'experts' and 'novices'. In the context of learning to write, this interest focuses not so much on the products of these two groups—although these are useful for characterising the differences between them—but more on the processes by which these two different kinds of groups achieve their products. For example, as we saw earlier, there is much debate and concern over how experts plan, write and revise compared with novices, and what kind of instruction is needed to help novices become experts.

Freedman et al. draw particular attention to the work of Bereiter and Scardamalia in this respect (See, for example, Bereiter & Scardamalia, 1987). This work is important because of the emphasis it places on teaching children (1) to monitor the acquisition of their own writing skills, and (2) to carry out more complex strategies than they would initially do on their own. The aim, in a nutshell, is to make the *learners* responsible for the evaluation of their learning rather than the teachers, and to teach learners how to do this.

Perhaps it might be helpful at this point to provide an example of how such teaching could take place. Scardamalia & Bereiter (1983), for instance, instructed 10- to 14-year-olds how to use a set of procedures for evaluating the effectiveness of their written compositions. The participants were asked to consider each sentence of something they had written and to check it, first against the list of evaluation statements, and second against the list of directives shown in Table 2.2. The aim of such an approach was to help the learners see what revision entailed, and then to make it a matter of routine. This example demonstrates the advantages of the procedure (it forces teachers to spell out what is meant by revision) and the disadvantages (not everyone will agree that this is what revision entails).

Table 2.2. Evaluative and directive phrases used to help 10- to 14-year-olds to revise in the study by Scardamalia and Bereiter.

Pupils check each of their sentences in turn, first against the list of evaluative phrases, and next against the list of directive statements.

Evaluative phrases

1. People won't see why this is important.
2. People may not believe this.
3. People won't be very interested in this part.
4. People may not understand what I mean here.
5. People will be interested in this part.
6. This is good.
7. This is a useful sentence.
8. I think this could be said more clearly.
9. I'm getting away from the main point.
10. Even I am confused about what I am trying to say.

Directive phrases

1. I think I'll leave it this way.
2. I'd better give an example.
3. I'd better leave this part out.
4. I'd better cross this sentence out and say it a different way.
5. I'd better say more.
6. I'd better change the wording.

Bereiter and Scardamalia have experimented with giving other children similar sorts of prompts and consequential composing activities (See e.g. Bereiter and Scardamalia, 1987; Scardamalia et al., 1981). For example, when 9- to 11-year-olds were given appropriate prompts to help them write opinions, they produced more structured elements (reasons, examples, elaborations) than when they attempted to do the same task simply by talking. Similarly, when children were given a choice of 'ending sentences' to strive for in their compositions, this was found to be more useful than a choice of 'sentence openers'. In another study, children were encouraged to list 'relevant' words as a part of their planning process. Children adopting this technique doubled the length of their essay and tripled their use of unusual words.

There have now been a fair number of such studies of learners improving their abilities at writing by using such self-instructional strategies (e.g. see Beal, Garrod & Bonitatibus, 1990; Brown & Day, 1983; Graham and Harris, 1989; Wallace & Bott, 1989). And, as noted in the introduction to this textbook, what is of interest here is how such procedures can be furthered by use of computer technology. Studies by Daiute and Kruidenier (Chapter 4) and Graham and MacArthur (1988) show the possibilities.

More generally, the procedures described above take place in the context of studies that are concerned with enhancing children's metacognitive skills (ie their knowledge of their own, thinking processes). A number of researchers now see writing as a major tool for teaching learners to enhance their thinking processes. Freedman et al. write:

> Initially children's language and thinking is embedded in ongoing events. During the preschool and early school years, children's language becomes free of what they can see and manipulate and thus becomes a tool for thinking and referring to the present, the past, and the possible future. Recently, researchers have emphasized the contribution of schooling and written language to the freeing of both language and thinking from immediate experience. In school, written language, and much of oral language, exists apart from a familiar social and physical setting—such language is 'decontextualized.' Children must reason about meanings conveyed primarily through words alone.
>
> This ability to transform an experience into language and then think about it—analyze it, compare it to previous experiences, and, perhaps, reinterpret it—is seen as the heart of higher-level cognitive functioning by researchers and theorists who have significantly affected current views on both cognitive and linguistic growth. The goal of education must be, in part, a reflective human being who is capable of 'intellectual self-control' of monitoring ongoing thinking, stopping and giving pause, considering possibilities and alternate routes, of taking necessary steps to disentangle confusions and make sense. A concern with the development of such a reflective citizenry is evident in the current interest in metacognitive skills, of individuals' knowledge about and control of their own thinking (Freedman et al., 1987; references omitted).

Several writers have commented on how writing aids thinking (e.g. see Langer & Applebee, 1987; Wason, 1970, 1980). Writers, with their text in front of them as they produce it, are able to reorganise it, to clarify what it is that they are trying to say, and sometimes to recognise that what they started out by saying is clearly no longer appropriate and needs changing. Debates concerning the roles of reading, writing and speech in the development of thinking skills have, according to Freedman et al., formed the impetus for the current enthusiasms for 'writing across the curriculum' programmes in our schools (e.g. see Fulwiler & Young 1982; Gray, 1988) and the processes-approach to instruction described earlier.

Finally, we should note that writing, even though it is often a solitary activity, does not take place in a social vacuum. Children learn from communication with their peers, parents, other adults and teachers: they watch and listen to a variety of media, each of which has its accompanying text. The recognition of this simple fact has led to an explosion of research work on the social nature of writing (see e.g. Freedman et al) and it permeates to some extent the chapters presented in this textbook. Studies of group authoring via computer exploit this social process (e.g. see Barrett, 1989a; Irish and Trigg, 1989; Rada et al., 1989). More general text books in this area are those of Barrett, 1989b; Goody, 1968; Heath, 1983; and Scribner & Cole, 1981. As noted earlier, this aspect of writing is reviewed in greater detail elsewhere by Freedman et al. under the heading 'The Uses of Writing'.

Evaluating the quality of writing

It is not a simple matter to evaluate the quality of a piece of writing, or to decide whether or not one piece of writing is better than another. Indeed, the temptation is to think that if we have re-written something then it must be better than it was before. But, if we are to move beyond opinion and value judgement, and try to assess whether or not, for example, computer technology makes a difference to how well people write, then we need to collect and analyse some data.

As we shall see later in the book, there have been many attempts to evaluate the effects of writing with word processors in our educational systems. Some of the early studies were limited because, for example, of the small number of computers available in one location, the small sample sizes available, and the brevity and quality of the intervention programmes. However, later studies, which remove some of these deficiencies, still seem to produce equivocal results (see Chapter 5).

Dunn and Reay (1989), in reviewing some of these later studies, point to the difficulties of making comparison studies. They emphasise, in particular, that investigators need to equate keyboard speeds in the experimental groups with handwriting speeds in the control groups before drawing conclusions. In their study, for example, Dunn and Reay found no significant differences between the quality of the writing of their 12- and 13-year-olds using word processors and those using handwritten composition when keyboard skills were ignored. However, when keyboard skills were taken into account, those writers who were adept (i.e. faster on the keyboard than by hand) did better on measures on writing quality, and those writers who were less competent (i.e. slower on the keyboard than by hand) did worse. These results may not seem surprising, but they point to the fact that in carrying out comparison studies (or indeed studies of any kind) one has to collect appropriate data.

The comparison study seems to be the *sine qua non* when it comes to evaluating new methods and techniques in education. A great deal of energy was expended, for example, in assessing the value of educational television programmes and computer assisted learning using this approach. However, once it has been determined that certain media accomplish some things rather well and some things rather badly (compared with human instruction) then the focus of evaluation usually turns to examining how one can make a more effective use of the medium concerned. It may be that we have not yet fully arrived at this stage in assessing the effects of computers on writing—although the current interest in writing processes may mean that we shall arrive at this position rather more quickly than was the case with other media.

Of course, assessing the effects of different processes on writing quality is not a simple matter—and it probably can never be done perfectly. My purpose in this final section of the chapter is to indicate the variety of tools available to researchers in this respect. The aim is to classify them in some way—to see what is available and to discuss the varieties and constraints that exist.

There seem to be at least three dimensions along which one can classify an evaluation study. First of all, as we saw in Table 2.1, one can categorise writing in terms of *different types*. In Table 2.1, it will be recalled, there are many different kinds of writing. Clearly, tools appropriate for evaluating one kind of writing may not be appropriate for another.

Secondly, we can also categorise *research methods* into different types or genres. Jaeger (1988), for instance, distinguishes between experimental (comparison) methods; quasi-experimental methods; survey methods; case-study methods, ethnographic methods; historical methods; and philosophical methods. The authors in Jaeger's textbook discuss the advantages and disadvantages of each in detail. As Shulman (1988) puts it: 'What distinguishes the methods from each other, by virtue of their contrasting disciplinary roots, is not only the procedures that they employ but the very types of questions they try to raise.' So, the kind of methods used in an evaluation study will depend upon the questions being asked. If we want to find out how teaching methods have changed, a survey method may be appropriate; if we want to compare the efficacy of different methods of instruction then a quasi-experimental procedure may be best; and, if we want to explore the feelings of participants, and underlying social and political issues, then we might opt for an ethnographic approach.

Bereiter & Scardamalia (1987) make a similar point in the context of research on writing when they outline six levels of enquiry. Table 2.3 shows how research at different levels poses different kinds of questions, and utilises different methods of approach. Bereiter & Scardamalia conceive of the different levels each having advantages and disadvantages, but all feeding into each other.

Thirdly, we can consider the variety of *measures* that might be appropriate for one or more of these research levels. For example, we can distinguish between a number of such measures as follows:

1. *Mechanical counting*. Here one can count, preferably by machine, the basic features of text (e.g. the number of key-strokes, the length of the product, etc.). The difficulties arise with how one interprets the results obtained. Faigley et al. (1985), for instance, comment, 'simple indexes of error or syntax have proven to be of little value for understanding the production skills of adult writers.'
2. *Classification*. This method again involves counting, but the data are classified in some way and recorded in categories. Daiute and Kruidenier, for instance (in Chapter 5), use a category system provided by Faigley and Witte for analysing pupils' revisions to text. Here there may be problems of overlapping categories, or of trying to decide whether a particular response is to be placed in, say, category *a* or category *b*. Some category systems may be easy to use, and some more difficult. Data from protocol analyses, and from interviews, for example, may be less easy to categorise than data from questionnaires. (The use of independent judges, however, is a useful way of examining or testing the consistency of category sys tems.)
3. *Subjective judgements*. A common method of evaluating writing is to use subjective judgements. In the world of essay-writing, for example, there is a considerable literature on the value of using holistic judgements to evaluate the quality of essays, as opposed to using more detailed marking schemes (see Huot, 1990). Some evidence suggests that holistic marking can be just as effective, and far less time-consuming, when the markers are

Table 2.3. Six levels of enquiry in research on writing - as outlined by Bereiter & Scardamalia.

Level	*Characteristic Questions*	*Typical methods*
Level 1: Reflective Inquiry	What is the nature of this phenomenon? What are the problems? What do the data mean?	Informal observation Introspection Literature review Discussion, argument, private reflection
Level 2: Empirical variable testing	Is this assumption correct? What is the relation between x and y?	Factorial analysis of variance Correlation analysis Surveys Coding of compositions
Level 3: Text analysis	What makes this text seem the way it does? What rules could the writer be following?	Error analysis Story grammar analysis Thematic analysis
Level 4: Process description	What is the writer thinking? What pattern or system is revealed in the writer's thoughts while composing?	Thinking aloud protocols Clinical-experimental interviews Retrospective reports Videotape recordings
Level 5: Theory-embedded experimentation	What is the nature of the cognitive system responsible for these observations? Which process model is right?	Experimental procedures tailored to questions Chronometry Interference
Level 6: Simulation	How does the cognitive mechanism work? What range of natural variations can the model account for? What remains to be accounted for?	Computer simulation Simulation by intervention

Table reproduced from Research on Writing: Principles and Methods, edited by Peter Mosenthal, Lynne Tamor and Sean A. Walmsley.

appropriately trained (Cooper 1977). One way to aid subjective judgement is to use some form of rating scales. Bailey (1988), for instance, describes numerous scales for assessing the quality of handwriting. And Brand, to take another example, has provided scales for assessing one's emotions whilst writing (see Brand & Leckie, 1988; Brand & Powell, 1986). People's ability to use such scales for assessing writing may be improved by repeated practice and by training. Clearer data, too, are sometimes obtained by averaging the data provided by several judges. Klare (1976), for instance, showed that the average judgement of 56 professional writers better reflected the difficulty of five text passages than did their individual judgements.

4. *Measures of efficiency.* Another common method of evaluating a piece of writing is to try and assess its efficiency at its purported task. In many studies of instructional writing, for instance, researchers are interested in assessing the effects of different versions of a text on the recipients—on their ability to understand the text, to recall salient points, to retrieve information from it, and perhaps to follow it in some way by using it to carry out a particular procedure. Wright (1987) illustrates how, in this connection, the criteria and the tools used for evaluating technical writing are very different from those used with narrative prose. Wright points out that with instructional and technical text there is a particular emphasis on obtaining feedback to see if the text carries out its purported function—and to improve it if it does not.
5. *Costs.* Finally, we may note that yet another way of evaluating efficiency might be to consider issues of cost-effectiveness. None of the chapters in this text devote themselves to this matter, and it is not an issue that is widely discussed in this context. Nonetheless, costs are an important consideration in the production of many kinds of instructional and informational texts. The difficulties of preparing cost-estimates, comparing the costs of different methods, and calculating direct and indirect cost-benefits are formidable, but such difficulties should not prevent people from attempting such measures in appropriate situations.

This description of five different kinds of measure has necessarily been over simple. My aim has been to show how one might classify approaches to evaluating the quality of writing in terms of a three-dimensional matrix: text types x research methods x different measures. It is perhaps not too surprising then to find that we cannot reach an overall simple conclusion concerning such questions as, does technology improve writing? Different investigators have used different cells in this matrix to tackle the question, and there is very little agreement as to what constitutes the best approach. And, we need to remember that for each cell, as pointed out by Dunn and Reay (1989), there is the question of whether the choice of text type, method or measure is entirely appropriate.

References

Applebee, A. N. (1986). Problems in process approaches: towards a reconceptualisation of process instruction. In Petrosky, A. & Bartholomae, D. (Eds.) *The Teaching of Writing: Eighty-Fifth Year Book of the National Society for the Study of Education*. Chicago, Ill.: University of Chicago Press.

Baer, V. E. H. (1988). Computers as composition tools: a case study of student attitudes. *Journal of Computer-Based Instruction*, 15, 4, 144–148.

Bailey, C. A. (1988). Handwriting, ergonomics, assessment and instruction. *British Journal of Special Education*, 15, 2, 65–71.

Barrett, E. (1989a). Textual intervention, collaboration, and the online environment. In Barrett, E. (Ed.) *op. cit.*

Barrett, E. (Ed.) (1989b) *The Society of Text*. Cambridge, Mass.: MIT Press.

Beal, C. R., Garrod, A. C. & Bonitatibus, G. J. (1990). Fostering children's revision skills through training in comprehension monitoring. *Journal of Educational Psychology*, 82, 2, 275–280

Beard R. (1984). *Children's Writing in the Primary School.* London: Hodder & Stoughton.

Bennett, N., Andrae, J., Hegarty, P. & Wade, B. (1980). *Open Plan Schools: Teaching, Curriculum Design*. Windsor: NFER.

Bereiter, C. & Scardamalia, M. (1987). *The Psychology of Written Composition*. Hillsdale, NJ: Erlbaum.

Bissex, G. L. (1980). *GNYS AT WRK: A Child Learns to Read and Write*. Cambridge, Mass.: Harvard University Press.

Bizzel, P. (1986). Composing processes: an overview. In Petrosky, A. & Bartholomae, D. (Eds.) *The Teaching of Writing: Eighty-Fifth Year Book of the National Society for the Study of Education*. Chicago, Ill.: University of Chicago Press.

Brand, A. G. & Leckie, P. A. (1988). The emotions of professional writers. *Journal of Psychology*, 122, 5, 421–439.

Brand, A. G. & Powell, J. L. (1986). Emotions and the writing process: a description of apprentice writers. *Journal of Educational Research*, 79, 5, 280–285.

Britton, J. (1972). *Language and Learning*. Harmondsworth, Penguin.

Britton, J., Burgess, A., Martin, N., McLeod, A. & Rosen, H. (1975). *The Development of Writing Abilities: 11–18*. London: Macmillan Education.

Brown, A. L. & Day, J. D. (1983). Macro-rules for summarizing texts: the development of expertise. *Journal of Verbal Learning and Verbal Behavior*, 22, 1–14.

Chomsky, C. (1971). Write first, read later. *Childhood Education*, March, 296–299.

Clay, M. (1975). *What Did I Write?* Auckland: Heinemann.

Cooper, R. C. (1977). Holistic evaluation of writing. In Cooper, R. C. & O'Dell, L. (Eds.) *Evaluating Writing: Describing, Measuring, Judging*. National Council of Teachers of English, 111 Kenyon Road, Urbana, Illinois 61801, USA.

DES (1978). *Primary Education in England*. London: HMSO.

DES (1982). *Education 5 to 9*. London: HMSO.

Dunn, B. & Reay, D. (1989). Word processing and the keyboard: comparative effects of transciption on achievement. *Journal of Educational Research*, 82, 4, 237–245.

Durst, R. K. & Newell, G. E. (1989). The uses of function: James Britton's category system and research on writing. *Review of Educational Research*, 59, 4, 375–394.

Elbow, P. (1973). *Writing Without Teachers*. New York: Oxford University Press.

Emig, J. (1971). *The Composing Processes of Twelfth Graders*. National Council of Teachers of English, 1111 Kenyon Road, Urbana, Illinois 61801, USA.

Faigley, L., Cherry, R. D., Joliffe, D. A. & Skinner, A. (1985). *Assessing Writer's Knowledge and Processes of Composing*. Norwood, NJ: Ablex.

Farr, M. (Ed.) (1985). *Advances in Writing Research: Vol. 1. Children's Early Writing Development*. Norwood, NJ: Ablex.

Fitzgerald, J. (1987). Research on revision in writing. *Review of Educational Research*, 57, 4, 481–506.

Francis, H. (1990). Learning to read and write. In Entwistle, N. (Ed.) *Handbook of Educational Ideas and Practices*. London: Routledge.

Freedman, S. W., Dyson, A. H., Flower, L. & Chafe, W. (1987). Research in writing: past, present and future. *Technical Report No. 1*. Center for the Study of Writing, University of California, Berkeley, CA 94720, USA.

Fulwiler, T. & Young, A. (Eds.) (1982). *Language Connection: Writing and Reading Across the Curriculum*. National Council of Teachers of English, 1111 Kenyon Road, Urbana, Illinois, 61801, USA.

Galton, M., Simon, B. & Cross, P. (1980). *Inside the Primary Classroom*. London: Routledge & Kegan Paul.

Glynn, S. M., Britton, B. K., Muth, K. D. & Dogan, N. (1982). Writing and revising persuasive documents: cognitive demands. *Journal of Educational Psychology* 74, 557–567.

Goody, J. (1968). *Literacy in Traditional Societies.* Cambridge: Cambridge University Press.

Graham, S. & Harris, K. R. (1989). Improving learning disabled students skills at composing essays: self-instructional strategy training. *Exceptional Children* 56, 3, 201–214.

Graham, S. & MacArthur, C. A. (1988). Improving learning disabled students' study skills at revising essays produced on a word processor and self-instructional strategy training. *Journal of Special Education*, 22, 135–152.

Graves, D. H. (1983). *Writing: Teachers and Children at Work.* Exeter, NH: Heinemann.

Graves, D. H. (1984). *A Researcher Learns to Write.* Exeter, NH: Heinemann.

Gray, D. (1988). Writing across the college curriculum. *Phi Delta Kappan*, 69, 10, 729–733.

Gubb, J., Gorman, T. & Price, E. (1987). *The Study of Written Composition in England and Wales.* Windsor: NFER-Nelson.

Hayes, J. R. & Flower, L. (1980a). Identifying the organisation of writing processes. In Gregg, L. W. & Steinberg, E. R. (Eds.) *Cognitive Processes in Writing.* Hillsdale, NJ: Erlsbaum.

Hayes, J. R. & Flower, L. (1980b). The dynamics of composing: making plans and juggling constraints. In Gregg, L. W. & Steinberg, E. R. (Eds.) *Cognitive Processes in Writing.* Hillsdale, NJ: Erlbaum.

Hayes, J. R. & Flower, L. (1983). Uncovering cognitive processes in writing: an introduction to protocol analysis. In Mosenthal, P., Tamor, L. & Walmsley, S. A. (Eds.) *Research on Writing: Principles and Methods.* New York: Longman.

Hayes, J. R. & Flower, L. S. (1986). Writing research and the writer. *American Psychologist*, 41, 10, 1106–1113.

Heath, S. B. (1983). *Ways with Words: Language, Life and Work in Communities and Classrooms.* Cambridge: Cambridge University Press.

Hillocks, G. R. (1984). What works in composition: a meta analysis of experimental treatment studies. *American Journal of Education*, 93, 133–170.

Hillocks, G. R. (1986). The writer's knowledge: theory, research and implications for practice. In Petrosky, A. & Bartholomae, D. (Eds.) *The Teaching of Writing: Eighty-Fifth Year Book of the National Society for the Study of Education.* Chicago, Ill.: University of Chicago Press.

Humes, A. (1983). Research on the composing process. *Review of Educational Research* 53, 201–16.

Huot, B. (1990). The literature of direct writing assessment: major concerns and prevailing trends. *Review of Educational Research*, 60, 2, 237–264

Irish, P. M. & Trigg, R. H. (1989). Supporting collaboration in hypermedia: issues and experiences. In Barrett, E. (Ed.) *op. cit.*

Jaeger, R. M. (Ed.) (1988). *Complementary Methods for Research in Education.* Washington: American Educational Research Association.

Juel, C. (1988). Learning to read and write: a longitudinal study of 54 children from first through fourth grades. *Journal of Educational Psychology*, 80, 4, 437–447.

Kellogg, R. T. (1988). Attentional overload and writing performance: effects of rough draft and outline strategies. *Journal of Experimental Psychology, Learning, Memory and Cognition*, 14, 355–365.

Klare, G. R. (1976). Judging readability. *Instructional Science*, 5, 1, 55–61.

Kurth, R. J. (1987). Using word processing to enhance revision strategies during student writing activities. *Educational Technology*, January, 13–19.

Langer, J. & Applebee, A.N. (1987). *How Writing Shapes Thinking.* National Council of Teachers of English, 1111 Kenyon Road, Urbana, Illinois, 61801, USA.

Maarse, F. J., van de Veerdonk, J. L. A., van der Linden, M. E. A. & Pranger-Moll, W. (1991). Handwriting training: computer aided tools for remedial teaching. In Wann, J., Wing, A. M. & Sovik, N. (Eds.) *The Development of Graphic Skills.* London: Academic Press.

Martlew, M. (Ed.) (1983). *The Psychology of Written Language: A Developmental Approach*. Chichester: Wiley.

Moore, O. K. (1962). *The Automated Responsive Environment*. New Haven: Yale University Press.

Murray, D. (1984a). *A Writer Teaches Writing* (2nd edition). Boston: Houghton-Mifflin.

Murray, D. (1984b). *Write to Learn*. New York: Holt, Rinehart & Winston.

Rada, R., Keith, B., Burgoigne, M. & Reid, I. (1989). Collaborative writing of text and hypertext. *Hypermedia*, 1, 2, 93–110.

Read, C. (1986). *Children's Creative Spelling*. London: Routledge & Kegan Paul.

Sarachno, O. N. (Ed.) (1990). Emergent Writing: Special Issue of *Early Childhood Development and Care*, 56, 1, 1–90.

Scardamalia, M., Bereiter, C. & Fillion, B. (1981). *Writing for Results: A Sourcebook of Consequential Composing Activities*. Toronto: Ontario Institute for Studies in Education.

Scardamalia, M. & Bereiter, C. (1983). The development of evaluative, diagnostic and remedial capabilities in children's composing. In Martlew, M. (Ed.) *op. cit.*

Schriver, K. A. (1986). Teaching writers to predict readers' comprehension problems with text. Paper available from the author, Department of English, Carnegie-Mellon University, Pittsburgh, Pennsylvania, USA.

Scribner, S. & Cole, M. (1981). *The Psychology of Literacy*. Cambridge, Mass.: Harvard University Press.

Shaver, J. P. (1990). Reliability and validity of measures of attitudes toward writing and toward writing with the computer. *Written Communication*, 7, 3, 375–392.

Shulman, L. J. (1988). Disciplines of inquiry in education: an overview. In Jaeger, R. M. (Ed.) *op. cit.*

Trueman, M. (1990). The effects of gender and computer experience on attitudes towards computers. *CORE (Collected Original Resources in Education)*, 14, 3, 29-43.

Wallace, G. W. & Bott, D. A. (1989). Statement pie: a strategy to improve the paragraph writing skills of adolescents with learning disabilities. *Journal of Learning Disabilities*, 22, 9, 541–543, 553.

Wason, P. C. (1970). On writing scientific papers. *Physics Bulletin*, 21, 407–8. Reprinted in J. Hartley (Ed.) (1980) *The Psychology of Written Communication*. London: Kogan Page.

Wason, P. C. (1980). Specific thoughts on the writing process. In Gregg, L. W. & Steinberg, E. R. (Eds.) *Cognitive Processes in Writing*. Hillsdale, NJ: Erlbaum.

White, J. (1986). *The Assessment of Writing: Pupils Aged 11 and 15*. Windsor: NFER-Nelson.

White, J. (1987). *Pupils' Attitudes to Writing*. Windsor: NFER-Nelson.

Wright, P. (1987). Writing technical information. In Rothkopf, E. Z. (Ed.) *Review of Research in Education, Vol. 14*. Washington: American Educational Research Association

Part II

Computers in the classroom

Chapter 3

A Case-Study of Collaborative Writing*

Colette Daiute

Observations of computer writing environments suggest that when children collaborate, they help each other learn effective techniques and forms of written communication, but close examination of collaborative texts indicates that the collaboration process is complex, requiring careful study if it is to lead to transfer of performance in collaboration activities to learning.

Most of the students in one study liked writing with partners: 'Each of us could have our own ideas and mix each of them to make the story better' (Bruce & Rubin, 1983, p.17). But, students also noted some problems: 'It's fun but sometimes they take too long and don't pay attention,' and 'Partners hog the computer' (Bruce & Rubin, 1983, p. 10). The change in children's writing does not yet confirm that collaborative writing enhances writing development: 'The changes in our posttest measures after a few months were not striking' (Reil, 1983, p.63). Another study reports significant changes in scores on texts written by students in classes that involved collaborative writing (Bruce & Rubin, 1983). The increases in scores were on persuasive and expository writing tasks but not narratives, suggesting that collaboration may have heightened students' sense of the reader, which is more important in persuasive and expository writing than in narrative writing. Both sets of pre- and posttest measures are from solo writing done in pencil or pen. No measures of solo or collaborative writing on the computer are offered. Several studies have suggested that regardless of the instrument, posttest writing scores increase at the end of a period in which students have written (Daiute, 1984; Levin, in press), but texts written on the computer are significantly different from those written in pen (Daiute, 1984; Levin, in press), so it is important to compare computer texts to pen texts when measuring change.

Case study on collaborative writing

A case study offered further information on practices and issues of collaborative writing. In this case study, two writers composed and revised in a variety of conditions: writing alone (solo writing), writing with a partner (collaborative writing), and writing with pencil versus on the computer. The purpose of this

---

* *Excerpt from Daiute, C. (1985) Issues in using computers to socialise the writing process,* Educational & Technology Communication Journal, *33, 41-50, © 1985 Association for Educational Communications and Technology, 1025 Vermont Avenue, NW, Washington DC 20005. Reproduced with kind permission of the author and AECT.*

study was to gain more information about collaborative writing and the role of the computer by comparing children's collaborative texts and solo texts. The study was also designed to offer information on the value of face-to-face collaborative learning versus collaborating remotely via a computer network, which would increase the children's need to read the text.

Two seven-year-old boys in a Cambridge, Massachusetts, school were identified by their teachers as 'good readers who were also interested in writing and computers'. The boys visited a laboratory at the Harvard University Graduate School of Education three times and wrote a total of eight texts. Each child wrote one text alone in pencil and one text alone on the computer (four texts); they also wrote four texts together—three on the computer and one in pencil. All the texts were written in response to story starts like 'Write a story about a bird who lives in a kid's lunchbox.' These themes were designed to elicit fantasy stories so the children could create a shared experience. Each story start included a main character in a setting with the potential for a narrative with a conflict. Young children tend to write narratives but have trouble resolving conflicts in their stories. The potential for conflicts suggested in the story starts was intended to provide material for collaborative discussions. These controls on the writing conditions make it possible to compare the children's texts.

The children, who previously knew each other only in passing, worked separately on the first stories they wrote in pen and on the computer. They each had about half an hour of training on basic features of the EDT word processing program on a PDP 1134, which each child used from terminals connected through phone wires. The children were familiar with using a computer keyboard as they had both had experience programming with the LOGO programming language (Papert, 1980) in school. The children wrote two of the collaborative texts from their terminals in different rooms, so they were collaborating through the text as it appeared on the computer screen. They took turns starting the collaborative stories. In each situation, the child had about five minutes to add to the story and then send the text to his partner to change or continue it. In the face-to-face collaborative writing condition on the computer, the children were told to make their own decisions about how to proceed.

The eight texts were analyzed for number of words in the text, number of words by each writer, syntactic complexity (in words per *t*-unit (Hunt, 1968), and through a discourse analysis measuring the development and follow-through among the dramatic elements in the stories.

This case study offers interesting information for future studies of collaborative writing with computers and with pencils. This study on collaborative writing with computers suggests that the computer facilitates physical aspects of collaboration at least for seven-year-olds who seem to have more trouble forming letters than pressing keys. The children wrote more words when they worked on the computer. The collaborative hand writing condition presented distractions: One child noted that it was too slow, and the other teased him for his sloppy handwriting. Another physical benefit of writing on the computer was that the children's contributions looked the same, which seemed to obscure the differences in authorship. Each child occasionally used the word processing editing commands on his own contributions, but since each was reluctant to change the other's writing, neither child used the word processing commands extensively across a text.

While theory indicates that writing in pairs makes writing easier, this study suggests that writing with a partner doesn't simplify the writing task, at least at first. Collaborative writing may be significantly different from solo writing, so children need time to practice and develop strategies for collaborating.

The texts and the collaborative experience indicate that although children enjoy working together, they may not have the social or cognitive prowess to collaborate effectively at first. An indication that collaborative writing situations may prove problematic was that the children expressed concerns over the development and ownership of the texts (Gerster, 1984). One boy didn't like one part of a story the other boy had contributed but didn't feel he should change an entry that wasn't his. Yet, this same child would interrupt his partner when he was tired of waiting for him to add a section or when he didn't agree with the story addition. Gerster suggests that such confusion over authorship in collaborative writing experiences requires explicit discussion between partners and establishment of rules. This study indicates, moreover, that such a negotiation over collaborative procedures would have to be guided by an adult, although basically left to the children. One child working on the collaborative story asked a researcher a question about the fate of a character in the story: 'Can I make him die?' The researcher said, 'You decide. It's your story.' 'No,' protested the child, 'It's *our* story.' This anecdote suggests that children can develop a sense of coauthorship, although they are somewhat confused at first.

Another factor influencing the collaboration was the children's differing personalities and work styles. One boy was deliberate in all his actions. he thought before writing, and read each sentence after he finished it. He scrutinized his partner's entries as closely as his own. The other boy was more impulsive—writing quickly, tending not to read the text. In attempts to work with his partner to solve a problem, the deliberate boy said 'What should we have happen now? What do you think David?' David responded in an impatient tone, 'Oh, have him die or something!' Frankie wanted to plan the story, while David simply wanted to get on with it.

The effects of personality on collaboration were also evident in the texts. Frankie wrote more words and more complex sentences than David on the solo pieces, but when they collaborated, David's entries were longer and more complex than Frankie's. Frankie's more deliberate work style was thwarted when collaborating with David, but David benefited from Frankie's influence. In our current studies, we are considering the children's work styles when we select collaboration pairs.

It may also be that seven-year-olds are too young to handle the cognitive demands of the collaborative writing task. The collaborative writing experience offers the writer the chance to get immediate responses from a reader, but writers may need time to know how to interpret and to use the response. At some points in development, collaboration may offer children models for internal problem solving, while at others they may not be able to use it well, or it may confuse them. Seven-year-olds may be too young to consider their partner's story goals, the technical aspects of recently acquired written language such as phoneme-grapheme correspondences, punctuation, and clear pronoun reference. They may be too young to take a reader's point of view even if another point of view is provided concretely by the partner. Frankie and David's collaboration indicated a readiness to create a complex text. The story,

'The Giant Who Lived in the T' (the Boston subway), shows that the children tried to weave in story elements introduced by the other writer, but they had trouble doing so. In this story, David tried to incorporate the information Frankie had introduced about the giant's sore foot.

> *David's contribution*: There once was a giant who lived in BOSTON in a VERY BIG house on 113 Washington St. *Frankie*: One day he had a sore foot and wanted to go to Kendall Sq. He knew about the T but he was too big so the little people decided to build a huge T. *David*: So he went in the huge T and burst it open because it was too small and he fell on the ground and got run over by a truck and threw it and he got arrested and burst open the jail because it was too small. He had to use his left foot because of his sore foot. Then he got scrunched into a heavy-duty titanium cell and he was awful squished in it.

The mention of the 'sore foot' in the second-to-the-last sentence is a noble attempt to weave in a story element, but it is awkward syntactically and thematically. The young writers didn't notice such problems even when they reread the texts, perhaps because they both were the writers and thus had difficulty taking a reader's point of view. In spite of attempts to write a coherent story, the children wrote more coherent stories when working alone than when collaborating. The children's solo stories were more smoothly shaped and worded. In their stories, the children tended to introduce the main character, set the character in context, and then develop a situation to the point of a conflict, which was a conclusion or was later resolved—albeit in an unsophisticated way.

To our surprise, an analysis of the children's discussions showed that they did not talk about the stories much in the face-to-face condition (Midkiff-Borunda, 1984). The face-to-face writing conditions on the computer and with pen followed the solo and the remote writing conditions, so the children may not have been comfortable talking although the researchers urged them to do so. In a current study, we are controlling the order of writing conditions. Of course, it is most likely that with practice, writers will develop strategies for social and cognitive aspects of collaboration, but the nature and speed of this acquisition will no doubt differ developmentally (Carey, in press).

The contrasts between the face-to-face collaborative texts and the remote ones written via the computer network offers preliminary information on another value of using computers. The children's collaborative texts written from different rooms but in the same computer space were slightly more coherent than the ones written face-to-face. This suggests that when working together via the computer network, they read the texts, something they may not have done when together and able to talk, even though their oral collaboration was not very elaborate.

The study suggests that social and cognitive factors of collaborative learning are complex. The computer offers young writers certain useful physical aids, and their attempts to weave a unified story reveal a sensitivity to story development, but the problems indicate new questions and issues for research and teaching.

Implications for research and teaching

The recent work on computers and socialization of writing suggests that the computer is a writing instrument that could significantly increase collaboration. Nevertheless, to be sure about the role of the computer, we need more detailed studies designed to reveal the differential effects of teaching philosophy, writing environment, writing assignment, students' work styles, and cognitive development. Detailed analytic and experimental evidence of writing on and off the computer will offer useful evidence.

This review also suggests some educational implications. Teachers need more time to plan the integration of the computer into the writing curriculum. Teachers need more time to integrate computers into the curriculum than to evaluate hardware and software, which has been a major focus of concern. Teachers need time to think about the ways in which the instructional technologies enhance (or change) their philosophy of writing, the curriculum, and their students' control over their own learning. Activities should be based on students' academic and personal needs rather than on the pressure to maximize the use of the computer. Thus, collaborative writing should be motivated by the goal of setting writing in communication contexts. The student pairings and cooperative procedures in such collaboration activities should be planned carefully. Most importantly, writers should continue to do a variety of writing activities alone and with instruments other than the computer to ensure that they are developing writing abilities rather than strategies to overcome artifacts of new instruments and procedures that have not been fully explored.

In conclusion, recent research has shown that computers can be used to change communication patterns in the classroom. This chapter has outlined several factors that are more important than the computer in stimulating the social changes. The teacher's philosophy about teaching writing, the writing activities, and the abilities and work styles of the student writers determine the ways in which the computer can be used to simplify the writing process. In any case, the new writing technology has been a catalyst to more significant changes in many writing classrooms. The job now is to understand clearly the causes so the results are no longer surprising.

References

Bruce, B., & Rubin, A. (1983). Report to the Office of Education on the QUILL Project. Washington, DC: Office of Education.

Carey, S. (in press). Are children fundamentally different kinds of thinkers and learners than adults? In Chipman, S. F., Segal, J. W. & Glaser, R. (Eds.), *Thinking and learning skills: Current research and open questions* (Vol. 2). Hillsdale, NJ: Erlbaum.

Daiute, C. (1984). *Rewriting, revising, and recopying*. Paper presented at the meeting of the American Educational Research Association, New Orleans.

Gerster, J. (1984). *Collaborative writing: Implications for the classroom*. (Paper for 'Computers and Writing'.) Cambridge, MA: Harvard University, Graduate School of Education.

Hunt, K. (1968). *Syntactic maturity at different grade levels*. Urbana, IL: National Council of Teachers of English.

Levin, J. (in press). From muktuk to jacuzzi. In Freedman, S. (Ed.), *The acquisition of written language: Revision and response*. Norwood, NJ: Ablex.

Midkiff-Borunda, S. (1984). *Analysis of the structural differences in oral and written narratives.* (Paper for 'Computers and Writing'.) Cambridge, MA: Harvard University, Graduate School of Education.

Papert, S. (1980). *Mindstorms.* New York: Basic Books.

Reil, M. (1983). Education and ecstasy: Computer chronicles of students writing together. *The Quarterly Newsletter of the Laboratory of Comparative Human Cognition*, 5(3), 59–67.

Suggested further reading

Barrett, E. (1989). Textual intervention, collaboration and the online environment. In Barrett, E. (Ed.) *The Society of Text.* Cambridge, Mass: MIT Press.

Cohen, M. & Riel, M. (1989). The effect of distant audiences on students' writing. *American Educational Research Journal*, 26, 2, 143–159.

Daiute, C. (1985). *Computers and Writing.* Reading, Mass.: Addison Wesley.

Daiute, C. (1988). Do 1 and 1 make 2? Patterns of influence by collaborative authors. *Written Communication*, 3, 3, 302–408.

Dipardo, A. & Freedman, S. W. (1988). Peer response groups in the writing classroom: theoretic foundations and new directions. *Review of Educational Research*, 58, 2, 119–149.

Fitzgerald, J. & Stamm, C. (1990). Effects of group conferences on first graders' revision in writing. *Written Communication*, 7, 1, 96–135.

Katstra, J., Tollefson, N. & Gilbert, E. (1987). The effects of peer evaluation on attitude towards writing and writing fluency of ninth grade students. *Journal of Educational Research*, 80, 3, 168–172.

Stevens, R. J. *et al.* (1987). Cooperative integrated reading and composition: two field experiments. *Reading Research Quarterly*, 22, 4, 433–454.

Underwood, G., McCaffrey, M. & Underwood, J. (1990). Gender differences in a co-operative computer-based language task. *Educational Research*, 32, 1, 44–49.

Webb, G. (1990). Case-study: developing writing with peer discussions and microcomputers. *Educational Training and Technology*, 27, 2, 209–215.

Chapter 4

A Self-Questioning Strategy to Increase Young Writers' Revising Processes*

Colette Daiute and John Kruidenier

Background and rationale

Studies of student writers (Bridwell, 1981; Perl, 1979; Calkins, 1980; Bereiter and Scardamalia, 1982) have shown that beginning writers do limited spontaneous revising. While experienced writers tend to revise extensively (Sommers, 1980), beginning writers tend to change only spelling or word choice in drafts. Revising requires a complex set of skills including identifying problems in texts and knowing how to remedy these problems (Hayes, 1984).

There are several reasons why young writers may not revise. Young writers assume that the texts they write say what they intended them to say, so revising seems unnecessary. They do not take the objective perspective that would help them say 'Ick, that's terrible,' about something they have written. From their point of view, they don't have to read what they have written because they know what the page says, and in the absence of a strategy for critical reading with the goal of improving the text, they simply don't know what to do when asked to revise. In addition to lacking revising strategies, young writers may not know what features to examine for insights about improving the text—making it more clear, concise, or interesting. Most students know to look for spelling mistakes, but they do not know how to use text features to identify syntactic, organizational, or rhetorical aspects of the text that may prove problematic for readers. Even if a student identifies a problem, he or she may not know how to correct it. Research has just begun to identify in detail these specific revision processes of identifying and correcting problems (Hayes, 1984).

Writing instruction often focuses on features of good and bad texts, but a list of features does not provide a strategy for reading, critiquing, and improving one's own work. Strategies that help writers take the position of a potential reader seem to be more useful for identifying and correcting problems in texts. This study explored the strategy of inner dialoguing about one's own writing as a method to help students interact with their texts more objectively and to revise more.

---

* *Reprinted from* Applied Linguistics, *1985, 6, 307-318, © Cambridge University Press. Reprinted by kind permission of the authors and Cambridge University Press.*

Experienced writers often have a strong sense of inner dialogue about their texts (Welty, 1983). This inner dialogue about the content, organization, clarity, and mechanics of a text is not always conscious or explicit, and it is rarely audible. As a writer, I find that I talk to myself most when I am having trouble composing a text or figuring out why some section doesn't sound right. For example, I might say to myself, 'Just what is this paragraph saying? What's my main point?' or 'This isn't convincing enough. What have I left out?'

In attempts to help young writers take more control over the writing process, teachers and researchers have recently begun to present writing instruction in the context of conversational dialogue. These researchers have observed that even young children can become more fluent writers (Graves, 1983) when they discuss their texts with peers. When young authors talk to their readers, they seem to be able to revise and expand their texts, presumably because they have had the benefit of taking an objective point of view about their writing. Relying on readers to supply the objective point of view is helpful, but writers may benefit more from learning how to ask themselves questions and learning the kinds of questions to ask. Since developing into a mature writer involves learning to be the reader as well as the writer, we need to understand how children internalize the social process of conferencing to autonomous inner dialoguing.

Research has shown that encouraging youngsters to use explicit self-monitoring strategies can help them improve their performance on memory and reading comprehension tasks (Brown et al., 1983). Children can also control their writing processes more when they engage in a variety of explicit planning and analysis activities (Bereiter & Scardamalia, 1982).

Research on procedural facilitation in writing (Bereiter & Scardamalia, 1982) has suggested that providing young writers with prompts and checklists to guide composing and revising activities can help students expand their writing processes. Researchers have observed that students will compose longer texts when they refer to questions such as 'Write more,' and they will revise more when they refer to questions such as 'Will this sentence be clear?' (Bereiter & Scardamalia, 1982). Studies on procedural facilitation suggest that young writers can benefit from analytical procedures, but there have been no detailed analyses of revising in relation to self-questioning strategy. We need more studies to show the relationship between inner dialogue and revising. The studies done thus far have involved providing students with lists of questions and features of writing. Another unexplored issue in these studies is that consulting prompts on notecards and checklists may present additional cognitive burdens on young writers and thus make it difficult for them to focus on analysis of their texts or to model conversational processes. The computer can be an interactive writing tool, but there has been little research on the effectiveness of prompting on the computer, which could be programmed to provide models for self-questioning.

The study reported in this chapter is grounded in recent cognitive-developmental research on the strategic and executive control of cognitive behavior (Brown et al., 1983; Flavell, 1976). The study extends the prior research on the use of prompting by exploring the strategy of conferring with oneself rather than with peers and by examining the relationship between an inner dialogue strategy modeled with the computer and revising processes. The goal of this study was to examine changes in students' revising strategies after they used a set of self-posed questions to guide their evaluation of texts. The assumption of this study was that if writers in junior high school developed the strategy of talking to themselves about a variety of text features that are often involved in revisions, they would revise more and in different ways. The practical import of using a self-questioning strategy is to manage efficiently the psycholinguistic processes of reading as well as writing one's texts. Another important part of this study was to explore prompting on the computer as a model for inner dialogue about the text.

Prompts on the computer can be presented one at a time *at the writer's request* in the context of the text any time during or after composing. Rather than having to shuffle through a set of questions on note cards and shift attention between the selected question and the text, the students can view questions of their choice under a section of text. While considering each question-prompt, the writer can use word processing functions to make changes. The word processing program simplifies the process of making changes. An assumption in this study was that beginning writers would need the guidance of prompting to benefit from the word processing features.

Hypothesis

The hypothesis in this study was that students who referred to question-prompts would revise more than students who used only a word processing program. Prior work (Brown et al., 1983; Daiute, in press) suggested, moreover, that students having difficulties learning to write would benefit greatly from guided self-questioning strategies. The program used in this study provides a strategy and general set of features subjects can use when examining drafts. This examination process was intended to engage them in reading their drafts more closely. Because the expected outcome of the study was that the students who used the prompting would revise more extensively than the no-prompting group, our primary measures of prompting effects were an analysis of the number, types, and meaningfulness of revisions.

Method

Subjects and setting

To test this hypothesis, we provided one group of subjects with a prompting program and a word processing program to use over a five month period, while a corresponding group of subjects who had the same writing teacher and the same writing instruction used only the word processing program.

The subjects in this study were students from 11 to 16 years old in a New York City public junior high school that served a lower middle class multi-ethnic neighborhood. The subjects placed in the writing classes had scored in the middle range for their grades on the California Achievement Test, which assigns grade-level reading scores. Most of the subjects in the study scored at grade level on the reading test, but informal comparisons suggested their writing skills were slightly weaker than those of students in the same grade in suburban public schools or private schools in the city.

The study was run in collaboration with a teacher who taught writing to seventh, eighth, and ninth graders. In his classroom, the teacher had eight Apple II plus microcomputers, two printers, and the software provided for this study: a touch typing program, a word processing program, and a program to prompt revising. During the two years prior to the experiment, the teacher had worked with the principal investigator of this study to learn how to use the computer for writing and to integrate the use of the computer into the curriculum and classroom environment (Daiute, 1981, 1982). This teacher had also been trained in the process writing approach (Graves, 1983; Calkins, 1983), so he encouraged peer conferencing about texts and revising. The classroom was run as a workshop in which students wrote both with pen and the computer. The teacher gave group lessons on writing techniques such as using anecdotes, developing paragraphs, and increasing vocabulary, but most of the class time was devoted to writing and discussing texts. Each student wrote certain assignments on the microcomputer, for at least one hour a week, and worked alone or with peers using pen on other writing and vocabulary activities. As classroom activities, the students wrote one 'book' on the computer about their group of friends and another 'book' in pen about a person they admired. These assignments were intended to guide the students to write about their own experiences and to generalize beyond their personal perspectives about the community in which they lived. The writing samples used for analysis in this study involved similar personal topics; they were composed during class time, using the same general procedure students followed for their other writing activities: producing a draft which was revised a few days later.

Twenty-six students in one seventh and one ninth grade class were randomly assigned to be the experimental subjects. Thirty-one students in corresponding grades with comparable reading scores and the same writing teacher were assigned to be the control group. The teaching method and the experimental procedures were the same in the experimental and control classes, except that in the experimental classes the word processing program used in the classroom was augmented with a self-questioning prompting program, which was introduced in January and continued to be used for about three months prior to the collection of the computer writing sample.

Writing and prompting tools

All writers in this study used a computer word processing program, and the experimental group also used a revision prompting program that was added to the word processor. Writers can call the revision prompting program (Daiute, 1981, 1982, 1983) from the word processing program when they feel they need help revising a text. The program suggests a self-questioning revising strategy and a set of general text features that are useful to examine when revising.

The prompting program offers a selection of twenty-two questions, analyses, and suggestions in relation to the text the writer has been working on. Some of the prompts are questions guiding the writer to think about content, organization, and wording. For example, if the writer selects the 'point' option, the prompt 'Does this paragraph state a clear point?' appears at the bottom of the screen under the paragraph the writer has composed. If the writer indicates by pressing 'y' that the paragraph has a point, the program presents the 'point' prompt under the next paragraph. If the writer indicates that the paragraph doesn't have a point, the subsequent prompt suggests that the writer reread and improve the paragraph. Since the text remains in the word processing program, the writer can make changes of any nature at any time. In addition to question prompts, the program offers several calculations and pattern-matching analyses with prompts. For example, words in the text that appear on a list of vague words such as 'stuff' and 'thing' are highlighted with the prompt 'The highlighted words may not be clear. Can you use more specific words?' There are prompts about the text completeness, clarity, organization, conciseness, sentence structure, and punctuation. Table 4.1 lists the categories and selected prompts available in the system.

Table 4.1. Options for checking

Completeness	*Clarity*	*Organization*
*point development	coherence reference	guide words summary
sentence structure	conciseness	punctuation
long sentences short sentences	empty words *vague words	commas

Sample prompts:

*Point:
Does this paragraph have a clear point?
*Vague words:
Do the highlighted words add meaning to your text?

The features and prompts were intended to make sense to students who had not had specific training in writing and revising. For example, considering the point of the paragraph might make sense to an eleven-year-old even if the student has not been taught to identify topic sentences. Most of the features were as commonsensical as this one. All of the major text features—completeness, clarity, organization, sentence structure, conciseness, and punctuation—were discussed by the teacher throughout the year in both the experimental and control classes. Although the program was designed primarily to provide a way of engaging writers in reading their texts, the prompts also offer a set of text features to consider. Each prompt also makes general suggestions about how one could improve a text, thus encouraging the student to make a change after identifying a problem.

Collection of writing samples

The instructions for all of the writing samples analyzed were of the following type: 'Write a letter to someone who lives far away from New York City. Explain why "New York City is a place of opportunity" or "New York City is not a place of opportunity". You have 15 minutes to write a draft. You will have time to revise in a few days.' The specific topic of the letter about New York City differed slightly in each sample; for example, 'New York City is a fun place to live' or

'New York City is not a fun place to live,' was used in the pretest. Like the other writing activities in the course, these parallel tasks drew on students' experiences and required that they generalize, but were not too demanding in terms of content or organization. While the experimental sample taken in April was written on the computer, the January pretest and a June posttest were written in pen, with revisions noted in a different color pen and then recopied.

Observations on the average 'composing burst' (Bereiter & Scardamalia, 1982) indicated that 15 minutes would be sufficient for writers in middle school to compose a letter on a personal topic. The subjects had a full 45-minute class period for revising, which occurred two to five days after the writing of the original drafts.

Revision analysis

The 171 January, April, and June draft/revision pairs were analyzed for number and type of revision. A taxonomy of revision types based on Faigley and Witte (1980) offered categories for analyses of superficial and meaningful additions, deletions, substitutions, reorganizations, consolidations, and distributions. Table 4.2 shows the categories in the Faigley and Witte Analysis.

Table 4.2. A classification of revision changes (Faigley & Witte, 1980)

I. Surface Changes
- *A. Formal Changes*
  1. Spelling
  2. Tense, Number, and Modality
  3. Abbreviation
  4. Punctuation
  5. Format
     - A. Paragraph
     - B. Other
- *B. Meaning-Preserving Changes*
  1. Additions
  2. Deletions
  3. Substitutions
  4. Permutations
  5. Distributions
  6. Consolidations

II. Text-based Changes
- *A. Microstructure Changes*
  1. Additions
  2. Deletions
  3. Substitutions
  4. Permutations
  5. Distributions
  6. Consolidations
- *B. Macrostructure Changes*
  1. Additions
  2. Deletions
  3. Substitutions
  4. Permutations
  5. Distributions
  6. Consolidations

In this revision coding schema, additions, deletions, etc are coded for the extent to which they affect the meaning of the text. Meaning-preserving additions do not change the meaning of a text at all; such an addition would be the unnecessary repetition or paraphrasing of a fact or idea already expressed. Additions, deletions, etc. may affect the meaning of the text in two ways; they may affect the 'microstructure' of the text by altering details or the 'macrostructure' by altering the overall meaning and main points in the text. The sentence 'In New York, there are also many computer schools' is an example of such a locally-meaningful (microstructure) addition in a text that has already noted a list of educational opportunities in New York. Typical of locally-meaningful additions, this sentence adds details to an idea or topic already discussed in the text but does not expand the overall text content. An addition that affects the global meaning of the text, however, adds a new topic or point such as in the sentence 'New York City offers many educational opportunities' in a text that had included one paragraph with examples of cultural opportunities and one paragraph with examples of entertainment opportunities. Such a global (macrostructure) addition would affect the summary of the text content rather than adding locally relevant details. The Faigley and Witte revision taxonomy was supplemented with three categories (superficial, locally meaningful, and globally meaningful) of additions at the end of the text, since these occurred frequently in the sample and were not distinguished in the original taxonomy from intratextual additions.

Five coders with backgrounds in linguistics or psychology achieved at least 75% agreement on each of twenty text pairs scored in common. In addition, each revision categorization was checked for agreement by a second researcher. After being coded, the April revision data were analyzed in a one-way Analysis of Variance, with the revision prompting program as the factor. A one-way Analysis of Variance was used to isolate prompting/no prompting differences because there was no other comparable sample taken on the computer. As reported elsewhere (Daiute & Kruidenier, 1984), a multifactor ANOVA (repeated measures over January, April and June) was used to compare writing in pen, writing on the computer, and prompting. The analysis considered here is concerned only with the prompting factor.

Results and discussion

There were no significant differences between the experimental and control groups just before the experimental intervention on the relevant revision measures: total revisions, additions within the text, additions at the end of the text, superficial revisions, locally meaningful revisions, and globally meaningful revisions. Significant differences between the groups in the April sample occurred in both number and types of revisions (see Table 4.3). The incidence of revisions declined significantly from January to April for both groups of subjects. This decline is probably associated with the use of the computer as opposed to pen, which in the larger study (see Daiute & Kruidenier, 1984) was found to decrease revision rate overall. Nonetheless, it is striking that experience with the prompting program retarded the decline in the experimental group's revisions to a significant degree.

Table 4.3. Mean number of revisions per 100 words on the January and April writing samples

	January (pen sample)		**April (computer sample)**	
Group sample	*Experimental*	*Control*	*Experimental*	*Control*
Measures				
Total revisions	22.5	22.8	13.6	9.3
Additions within	4.7	5.3	3.0	1.3
Additions at end	.3	.6	1.6	2.0
Superficial revisions	2.8	3.3	2.0	0.7
Locally meaningful revisions	2.1	2.4	1.9	2.3
Globally meaningful revisions	.1	.2	0.7	0.3

The experimental group made more revisions than the control group

The 26 experimental subjects made significantly more ($p<.05$; $F[1,55] = 5.45$) revisions per 100 words (13.6) than the 31 control subjects did (9.3). This greater revision rate by the experimental subjects indicates that students who use prompting consider their draft texts critically and act on their reflections about the drafts.

The experimental group revised more interactively

Interactive revising (Calkins, 1980) can be distinguished from less sophisticated revision behavior in a number of ways. Interactive revising involves making changes in content rather than in punctuation or spelling. It involves working with the original draft rather than redrafting a text from memory, and it involves making changes and additions within the text rather than at the end, as is common among writers who feel that the text already reflects their knowledge and their best effort at expressing it. Interactive revising, which involves reading a text, critiquing, and reworking it, is rare among beginning writers of all ages (Calkins, 1980; Bridwell, 1981). The experimental subjects in this study, however, revised more interactively than did the control subjects.

The experimental group made significantly more ($F[1,55] = 5.6$) additions within the body of their draft texts (3.0) than did the control group (1.3). The experimental subjects' interactive revising is reflected in the patterns of text addition—the most frequent revising activity identified in this study.

Experimental subjects made more meaningful revisions

An analysis of the meaningfulness of additions (whether they augmented the text in detail or in major points) indicated that the experimental subjects made significantly more ($F[1,55] = 5.45$) additions of main points (0.7)—macrostructure changes (described above)—than did the control group (0.3). The frequency of these changes is small, but since this type of expansion is uncommon by writers of this age, it is notable that we found it at all let alone at a significant level. While only 1 of 31 control subjects made an expansion of main points within his text, five of 26 experimental subjects made such changes. Experimental subjects also made more superficial additions than the control subjects but the same number of locally meaningful additions. A qualitative analysis of the main point additions indicated that these additions were not tangential to the focus of the paper. The following draft/revision pair illustrates the nature of interactive revising by experimental subjects.

> *DRAFT*
>
> New York is with many opportunity because there are many places to live. There are also many plce to and god time there many fun places to go. We have important buildings to vist like the twin towers or the enpire state building and many other place New York also have many nice park to go to. There are much more many
>
> *REVISION*
>
> New York is full of many opportunitys because there are many places to live in. There are also many great jobs. New York is a very big. There are many nice places to vist. New York have important building and the twin towers and many more. New York has many tall skyscraper. The trains in New York are fast. The are many graet big nice parks.

This revision involves a relatively large number of both superficial and meaningful changes within the draft text. Even though the draft text and the revision seem immature and incomplete, this experimental subject's revision strategy is more interactive and at a higher level of meaning than one by a typical control subject, who adds to the ends of the texts.

An obvious question is whether differences in revising with computer prompting transfer to writing in general. The multifactor repeated measures analysis (Daiute & Kruidenier, 1984) indicates that effects related to where revisions occur (within or at the end of a text) and how they affect the meaning of the text (superficial, locally meaningful, or globally meaningful revisions) may carry over to the no-prompting conditions in pen. However, in this analysis the effects of prompting and writing instrument are confounded. Future studies will separate these effects and look specifically at transfer of the prompting intervention.

Interpretation

In summary, students who refer to question-prompts on the computer revise more often and more interactively than students who do not use prompts. There are several possible reasons why the prompts might have led to increased and closer revising of drafts. In addition to suggesting a self-prompting strategy, the prompts could have served to consolidate instruction on revising; they could have also served to focus the subjects' attention and thus make revising more possible. But, the data suggest most strongly that the prompts offered a model for self-prompting which engaged the subjects in a closer reading of their texts.

The prompts provide a range of text features to check for when revising. Our present results suggest that the specific features noted in the prompts were used by the subjects as a basis for text examination. Spelling correction, for example, which was not prompted, was not frequent by the experimental group, but it was by the control group. Thus, it seems that the nature of the prompts does have some influence on the subjects' revising activity, even when the prompts are general, as they were in this study. Since the students in both the experimental and control groups were involved in classroom instruction and discussion of all the text features that appeared in the prompting program, we do not attribute a teaching role to the program. Of course, because of their repeated use of the prompting program, the experimental subjects may have been better able to incorporate the classroom instruction into their writing processes. As subjects considered the prompts they would like to use, they were reminded of important features of a text such as completeness, clarity, and organization. The category labels on the prompt menu (Table 4.1) could have served to stimulate and organize the revising process.

The prompts could also have led to closer revising because they helped the writers focus their attention on specific sections and features of texts. One major difficulty facing writers is the need simultaneously to monitor and to engage in disparate activities in short-term memory (Daiute, 1981; Flower & Hayes, 1981). Writers in the present study chose prompts one at a time, and these prompts appeared in the context of the text. Some prompts isolated individual paragraphs, and others highlighted sections, sentences, or words. Such a selection process could have led to increased revision because it is directed and focused. Examining a text for one issue at a time may result in more revising than examining a text generally 'to make sure it is okay', as many students do. When even relatively inexperienced writers focus their attention closely on sections of text rather than on the entire text at one time, they may be better able to identify problems and attempt to make improvements.

The experimental subjects' interactive revising patterns suggest most strongly that the prompts engaged them in reading their texts even more closely and critically than did the control subjects. The fact that the experimental subjects made more changes within their texts suggests that they read them, and the fact that the control subjects made relatively few internal text changes but added long, somewhat repetitive sections to the ends of texts suggests that they only skimmed their drafts. The experimental subjects' pattern of inserting generalizations of details and new topics within the text rather than at the end suggests a closer reading of the text. If one stops to make a substantial revision, one is more likely to have read the text rather than skimmed it. The higher rate of globally meaningful revisions by the experimental subjects also indicates that more of them were engaged in closely reading their texts. A subject who adds new main topics to the text rather than only additional details must have read the text to evaluate the content and made fine distinctions between just adding words and adding words that would expand the meaning of the text.

Observations of the students indicated the typical pattern was to select prompts, read them, and then take time to consider them. Subjects were often heard disagreeing with a suggestion in a prompt and then reading their text aloud. Subjects' informal reports about the prompting process also indicated that the prompts drew them into reading their texts. One subject, for example, reported that he used the 'Long Sentences' feature most of the time because 'it got him to read long sentences where he often found problems' (Daiute, 1983). In future studies, we will attempt to separate the possible effects of repeated exposure to revision categories and attention focusing aspects of computer prompting from the strategic influence of self-questioning and closer reading.

Significance

This study extends explorations on the value of helping children develop and control their intellectual skills by suggesting cognitive self-monitoring strategies. These results add support to the research (Brown et al., 1983), suggesting that the use of explicit verbal strategies can help students take more control over their cognitive processes. This is a specific case of the use of self-question-prompts on the computer for increasing youngsters' control over their thinking and their writing as evidenced by increased and expanded revising activities. This study offers specific information on the importance of providing beginning writers with instruction on the conversational strategy of self-questioning. Self-question-prompts lead to closer revising of texts, suggesting that prompts engage writers in reading their texts objectively. When young writers are given even general suggestions to evaluate their texts, they have the ability to do so and to decide, at least sometimes, on the appropriate expansions to make.

This study also suggests that in teaching writing, we provide students with strategies and heuristics to use autonomously when revising. They need extra help in reading their texts critically, but the help can be subtle and writer-controlled. Interventions to help writers become objective about their texts have up to now been initiated and guided mostly by teachers or peers rather than by the writers themselves, who can—as this study suggests—become critical of their own writing.

Acknowledgements

The authors are grateful to the Spencer Foundation for supporting this four-year study. We also thank Tom Bever for his help on conceptual as well as procedural aspects of this study, Karen Brobst, Terry Tivnan, and Bob Fallonsbee for their help with the statistical analyses; Arthur Shield, Pegeen Wright, Sharon Liff, Karen Ann Kazer, Sandy Mazur, Janet Liff, and Nancy McManus for assisting in running the study and analyzing the many papers it produced.

References

Bereiter, C. & Scardamalia, M. (1982). From conversation to composition: the role of instruction in the developmental process. In Glaser, R. (Ed.), *Advances in instructional psychology*. Hillsdale, NJ.

Bridwell, L. S. (1981). Revising by student writers. *Research in the Teaching of English*.

Brown, A. L., Bransford, J. D., Ferrara, R. A. & Campione, J. C. (1983). Learning, remembering, and understanding. In Mussen, P. H. (Ed.), *The handbook of child psychology*. Vol 3. New York: Wiley.

Calkins, L. (1980). Children's re-writing strategies. *Research in the Teaching of English*, 331–341.

Calkins, L. M. (1983) *Lessons from a child: On the teaching and learning of writing*. Exeter, NH: Heineman.

Daiute, C. (1981). Psycholinguistic foundations of the writing process. *Research in the Teaching of English*, 5–22.

Daiute, C. (1981, 1982, 1983). *The effects of automatic prompting in young writers. Interim reports to the Spencer Foundation*.

Daiute, C. (1984). Can computers stimulate writers' inner dialogues? In Wresch, W. (Ed.), *A writer's tool: the computer in composition instruction*. Urbana: National Council of Teachers of English.

Daiute, C. (1984). Rewriting, revising, and recopying. Paper presented at the meeting of the AERA, New Orleans, April.

Daiute, C. (in press). Do writers talk to themselves? In Freedman, S. (Ed.), *The acquisition of written language*. Norwood, NJ: Ablex.

Daiute, C., & Kruidenier, J. (1984). *Strategies for reading one's own writing*. Unpublished manuscript.

Faigley, L. & Witte, S. (1980). Revision. Paper presented at the meeting of the NCTE, Cincinnati.

Flavell, J. H. (1976). Metacognitive aspects of problem solving. In Resnick, L. B. (Ed.), *The nature of intelligence*. Hillsdale, NJ: Erlbaum.

Flower, L. & Hayes, J. R. (1981). A cognitive process theory of writing. *College composition and communication*, 32, 365–388.

Graves, D. (1983). *Writing: Teachers and children at work*. Exeter, NH: Heinemann.

Hayes, J. R. (1984). Processes in revision. Paper presented at the meeting of the AERA, New Orleans.

Perl, S. (1979). The composing process of unskilled college writers. *Research in the Teaching of English*, 13, 317–336.

Sommers, N. (1980). Revision strategies of student writers and experienced writers. *College Composition and Communication*, 31, 378–388.

Welty, E. *(1983).One writer's beginnings*. Cambridge, MA: Harvard University Press.

Suggested further reading

Beal, C. R., Garrod, A. C. & Bonitatibus, G. R. (1990). Fostering children's revision skills through training in comprehension monitoring. *Journal of Educational Psychology*, 82, 2, 275–280.

Bereiter, C. & Scardamalia, M. (1987). *The Psychology of Written Composition*. Hillsdale, NJ: Erlbaum.

Brown, A. L. & Day, J. D. (1983). Macro-rules for summarising texts: the development of expertise. *Journal of Verbal Learning and Verbal Behavior*, 22, 1–14.

Cochran-Smith, M., Paris, C. L. & Kahn, J. L. (1991). *Learning to write differently: Beginning writers and word-processors*. Norwood, N. J.: Ablex.

Graham, S. & McArthur, C. (1988). Improving learning disabled students' skills at revising essays produced on a word-processor: self-instructional strategy training. *Journal of Special Education*, 22, 133–152.

Wallace, G. W. & Bott, D. A. (1989). Statement pie: a strategy to improve the skills of adolescents with learning disabilities. *Journal of Learning Disabilities*, 22, 9, 541–543, 553.

Chapter 5

Word Processing and the Teaching of Writing*

Susan M. Zvacek

Introduction

Communication is the process that builds and sustains our social, cultural, and political structures. As such, the importance of written communication as a means of recording and storing information vital to our heritage cannot be overemphasized. Our educational systems have been charged with preparing individuals to use written communication effectively.

Composition instruction at all educational levels is being revitalized to take advantage of many technological innovations available to support and enhance the writing process. This article will discuss how the microcomputer and word processor are being used in writing instruction. Research on the following topics will be reviewed:

1. how word processing influences the composition process,
2. the motivational aspects of word processing, and
3. how word processing affects the final product.

Examples of successful implementations of word processing will be reviewed, along with some practical issues and questions for future research efforts.

Word processing

Word processing is a procedure by which text may be entered into the computer's workspace to be edited, printed, saved, and later retrieved for further use. This article will discuss only those word processing programs that have these functions—disregarding programs with special writing features, such as spelling checkers and style editors. For the purposes of this article, word processing is defined as a software program that utilizes the computer as a tool to facilitate and support the composition process.

Teaching writing

A major shift has been taking place in composition instruction over the last several years. The trend toward 'process oriented' writing emphasizes the holistic, recursive nature of composition, rather than teaching writing as a

---

* *Reprinted from* Computers in Human Behavior, *1988, 4, 29-35, © Pergamon Press. Reprinted with kind permission of the author and Pergamon Press.*

neatly organized, linear process of discrete components. 'Word processing reinforces and enhances the dynamic, interactive, social nature of writing' (Parson, 1985, p.27).

Revision strategies

Proponents of word processing for composition instruction often cite revision strategies as the major advantage. The importance of revision to experienced writers cannot be underplayed, and word processing does indeed facilitate text revision, if for no other reason than the ease with which changes can be made without the drudgery of retyping or handwriting for a fresh paper copy.

Text revisions are often categorized by the type of change made in the document. A surface or microstructure revision is one in which the meaning of the text does not change, but a correction or modification such as spelling or punctuation is performed.

The other type of text revision includes those modifications that change the meaning or overall organization of the document. These macrostructure revisions are considered more meaningful in the composition process because they affect the content of the message received by the reader. Holistic, process-oriented writing instruction emphasizes these content revisions and encourages the beginning writer to concentrate on the meaning behind the words—whether on the screen or the printed page.

Researchers Bridwell & Duin (1985) have warned that word processing may actually encourage a preoccupation with surface revisions to the near-exclusion of revisions to meaning and overall structure of a document. Daiute (1985) also found that because word processing results in a polished look to the finished text, students may find themselves lured into concentrating mainly on modifications that make the document look good. Collier (1983) hypothesized, however, that a lack of content revisions by young writers could simply be due to their intellectual immaturity and inability to conceptualize complex revisions, and not the result of word processor usage.

For many composition instructors, word processing holds the key to increased revision (Bean, 1983; Daiute, 1986; Kelly, 1987; McAllister & Louth, 1987; Sommers, 1985), and students seem to require less encouragement to revise their writing efforts (Dalton & Watson, 1986; Kurth, 1987). (Only a few researchers—Harris, 1985 and Schanck, 1986, for example—have detected no significant differences in revising between groups using word processing and writing by hand.)

Several reasons have been suggested to account for an increase in revising. Probably the most obvious is the ease with which changes can be made. Given the choice of retyping or handwriting a long essay to modify one or two sentences and the option of turning in a less-than-perfect final product, most students would opt for the latter, feeling that the time and effort involved in revising the text would outweigh the potential benefits. With word processing, the physical, labor-intensive aspect of writing diminishes markedly.

The computer, by providing a polished-looking paper with each printing, may encourage students to produce a document worthy of clean copy. Lindemann and Willert (1985) found that high school students took pride in their work and its professional appearance seemed to 'engender a desire for perfection' (p. 49). It makes sense, intuitively, that if an essay is covered with erasures, strikeovers, and smudges, the content loses its appeal as well.

Another reason for the flexibility shown in text revision on the word processor was suggested by college writing students participating in a study by Collier (1983). The neutrality of text printed on the computer screen (the words look the same no matter who types them) may prevent writers from becoming 'infatuated' with their handwritten words, so that they more willingly experiment, modifying their writing to more closely represent their ideas.

Collaboration

Word processing reduces the solitary nature of the writing process, facilitating the display of partially completed manuscripts for 'conferencing' and group editing. Daiute (1985) found that 'children spontaneously share their writing' when computers are used in writing classes (p.18). Collaborative writing also increases for junior high students (Kane, 1983), high school students (Lindemann & Willert, 1985), and college students (Kelly, 1987; Kurth, 1987) when word processing is used. By networking computers together, or simply multiple loading word processor documents, files are easily shared, thus encouraging and fostering group efforts.

Motivation

Researchers citing the motivational aspects of using word processors often report this factor as an aside or note of interest, focusing predominantly on achievement. This emphasis on quantitative results mistakenly downplays the major influence that motivation has on student growth and development in writing.

Palmer, Dowd, and James (1984) working with elementary students, and Feldman (1984) working with college students, found that attitudes toward the use of word processing itself were overwhelmingly positive. For many students, the novelty of using a computer provides an impetus to write more often and utilize more of the features of the word processing software. Computers are still an innovation in most classrooms and the enjoyment from participating in a 'technological revolution' works to the student's academic advantage. How long this novelty effect will last remains to be seen.

A more relevant question, say some researchers, concerns student attitudes toward writing—not just working on the computer. In examining attitudes toward writing, students using word processing were more positive in their evaluations than were noncomputerized students for both Kurth (1987) and Sommers (1985). Moran (1983) and Teichman and Poris (1985) found that use of word processing reduced writing anxiety in students, creating a more positive attitude. Mike Sharples, in *Cognition, Computers and Creative Writing* (1985) provides a strong argument to explain why word processing builds positive feelings about writing.

> Children do realise the power of written language and they attempt to lay claim to it through graffiti, poems, diaries and stories. To be a writer is empowering, yet every word that a child forms on paper is a confirmation

> of inferiority. However carefully and neatly a child may write, the result is a poor substitute for adult typeface (p.10).

When students are equipped with the tools necessary to produce written works with a sense of credibility, they will take pride in their writing and want to improve it.

Writing quality

After considering revision strategies, collaborative activities, and motivational aspects, the question remains: does using word processors influence the quality of the writing produced? Results are mixed. A few researchers (e.g., Hawisher, 1986; Kurth, 1987) have discovered no significant differences in writing quality between word processing writers and 'traditional' writers. There are, however, a greater number of studies showing differences in the quality and quantity of writing when using the computer. Dalton and Watson (1986) found that junior high students in remedial language arts wrote significantly better when using a word processor. Teichman and Poris (1985) found, over a two-year study of 320 college freshmen, that the writing of word processing students was of better quality than nonword processing students. Etchison (1985) studied the work of 200 freshmen composition students and found that over a semester's time, those using word processing wrote essays of better quality and greater length. Experienced writers (college faculty, administrators, and system designers) produced longer and better-written documents when using word processing, according to Haas and Hayes (1986). From her work with elementary students, Jacoby (1984) implies that quality and document length are related due to the 'end of the page effect' that fails to occur in computerized writing. (Students often end their essays when they reach the bottom of the paper—something that doesn't happen on a computer screen.) Unfortunately, until a stronger body of rigorous research utilizing large sample sizes and replicable designs is built, conclusive evidence will remain elusive.

Issues in implementation

Practical applications

Suggestions for the appropriate implementation of word processing in composition instruction have begun to emerge from the burgeoning body of research literature of the past five years. Dalton and Watson (1986) and others have emphasized the need to teach word processing separately from writing. Once students attain proficiency at the keyboard, improvement of writing skills will grow unhindered by the distractions of making the machine 'behave'.

A second suggestion is that composition instructors may need to reassess traditional writing models. Selfe (1985) affirms that 'we can't continue to present our students with strategies designed for paper and pencil when we want them to experiment with the real power of the electronic pen' (p. 65). Composition instruction that exploits the computer's unique capabilities is needed.

Barriers

There are many barriers to the successful computerization of writing instruction. Lack of equipment is probably the major culprit, even though school purchases of microcomputers have increased dramatically in the last few years. Writing, unlike many instructional applications of computers, however, is not an efficient, step-by-step process for most students. Instead, they may need time in composing at the computer to reflect on their efforts and experiment with alternate structures. In order for students to fully integrate their writing strategies into a word processing model, large chunks of uninterrupted computer time are necessary—something not easily accomplished when a class of 25 students shares one computer, or when computers are housed in a lab with limited 'open' hours.

Disadvantages to word processor implementation cited by Sommers (1985) included the time taken away from writing for learning the word processing system, mechanical failures, and the loss of files due to operator error or system crashes.

Word processing may not be advantageous to every writer. Womble (1985) reported that some students had difficulty with large-scale revisions because only a portion of the text is available for viewing on the screen. These 'global' writers tended to edit on paper copies and then transfer the modifications onto their computer version later.

There are also writers who feel inhibited creatively when using the computer (Selfe, 1985). The physical act of writing their words on paper stimulates their thoughts. These 'tactile' writers will probably use the computer as a glorified typewriter, shunning the keyboard for the actual composing process.

Equity issues

One issue surrounding the implementation of word processing in composition involves equity of opportunity. Socio-economic factors heavily influence the degree of success in the language arts (Florio-Ruane & Dunn, 1985; Herrmann, 1987) and Watt (1983) suggests that the most likely candidates for writing improvement are children with computers at home, with well-educated, affluent parents.

Teachers need to be aware of their own biases and preconceived ideas about computer access and provide positive experiences for all students. These early experiences, says Stephen Marcus (1987), 'establish the foundation of what students (and faculty) think computers are for and *who* they think computers are for' (p. 134). Computers are well on the way to further widening the gap between the 'technology haves' and 'have-nots'.

Questions for future research

Based on this selective review of the research literature, several questions seem appropriate for future research. First, because of the short-term nature of almost all of the research (most of the aforementioned studies were conducted over one semester, or less), the impact of word processing over time is unknown. Does the novelty effect of composing at the computer wear off? If so, when? Would this have an adverse effect on a student's motivation for all aspects of writing? Studies examining the word processor's effects over time could have a significant impact on the nature of composition instruction.

Another area for long-term research would be the possible changes in writing models or strategies. Do writers using word processors develop strategies especially suited to the unique features of the computer? Do these strategies result in more efficient composition processes? How do writing models adapted to the computer influence the quality of the final documents? Again, long-term research of rigorous design and control of confounding variables would be necessary to test these research questions.

If, as suggested earlier, writing with a word processor is not for everyone, how can educators identify those students for whom the computer is, in fact, counter-productive, as well as those that will reap the greatest benefit? How might the growth and development of both types of writers be encouraged?

Word processing has become a permanent fixture in many composition classes and promises to expand even further. The use of computers in writing has gained acceptance among composition instructors, and the National Council of Teachers of English (NCTE) also endorses it, affirming that 'word processing supports the process model of writing instruction in ways that no other educational tool can' (Thomas, 1985, p. 2). The development of a strong research base in this field has begun and must be continued and built upon, adding to the accumulated store-house of theories and applications from both composition and computer research.

References

Bean, J. C. (1983). Computerized word processing as an aid to revision. *College Composition and Communication*, 34, 146–148.

Bridwell, L. & Duin, A. (1985). Looking in depth at writers: Computers as a writing medium and research tool. In Collins, J. L. & Sommers, E. A. (Eds.), *Writing on-line: Using computers in the teaching of writing* (pp. 115–121). Upper Montclair, NJ: Boynton/Cook.

Collier, R. M. (1983). The word processor and revision strategies. *College Composition and Communication*, 34(2), 149–155.

Daiute, C. (1985). *Writing and computers*. Reading, MA: Addison Wesley.

Daiute, C. (1986). Physical and cognitive factors in revising: Insights from studies with computers. *Research in the Teaching of English*, 20(2), 141–159.

Dalton, D. W. & Watson, J. F. (1986, January). *Word processing and the writing process: Enhancement or distraction?* Paper presented at the annual meeting of the Association for Educational Communications and Technology, Las Vegas, NV. (ERIC Document Reproduction Service No. ED 267 763).

Etchison, C. (1985). *A comparative study of the quality and syntax of compositions by first year college students using handwriting and word processing*. Unpublished manuscript. Glenville, WV: Glenville State College. (ERIC Document Reproduction Service No. ED 282 215).

Feldman, P. (1984). Personal computers in a writing course. *Perspectives in Computing*, 4(2), 4–9.

Florio-Ruane, S. & Dunn, S. (1985). *Teaching writing: Some perennial questions and some possible answers* (Occasional paper No. 85). Washington, DC: National Institute of Education (ERIC Document Reproduction Service No. 261 399).

Haas, C., & Hayes, J. R. (1986). *Pen and paper vs. the machine: Writers composing in hard copy and computer conditions*. CDC Technical Report No. 16. Pittsburgh, PA: Carnegie-Mellon University, Communication Design Center. (ERIC Document Reproduction Service No. ED 265 563).

Harris, J. (1985). Student writers and word processing: A preliminary evaluation. *College Composition and Communication*, 35(3), 323–330.

Hawisher, G. E. (1986, April). *The effects of word processing on the revision strategies of college students.* Paper presented at the annual meeting of the American Educational Research Association, San Francisco, CA. (ERIC Document Reproduction Service No. ED 268 546).

Herrmann, A. W. (1987). An ethnographic study of a high school writing class using computers: Marginal, technically proficient, and productive learners. In Gerrard, L. (Ed.) *Writing at century's end* (pp. 79–91). New York: Random House.

Jacoby, A. (1984, February). *Word processing with the elementary school student—A teaching and learning experience for both teachers and students.* Paper presented at the Spring Conference of the Delaware Valley Writing Council and Villanova University's English Department. Villanova, PA. (ERIC Document Reproduction Service No. ED 246 449).

Kane, J. H. (1983, April). *Computers for composing.* Paper presented at the annual meeting of the American Educational Research Association, Montreal, Canada. (ERIC Document Reproduction Service No. ED 230 978).

Kelly, E. (1987). Processing words and writing instructions: Revising and testing word processing instructions in an advanced technical writing class. In Gerrard, L. (Ed.), *Writing at century's end* (pp. 27–35). New York: Random House.

Kurth, R. J. (1987, April). *Word processing and composition revision strategies.* Paper presented at the annual meeting of the American Educational Research Association, Washington, DC. (ERIC Document Reproduction Service No. ED 283 195).

Lindemann, S. & Willert, J. (1985). Word processing in high school writing classes. In Collins, J. L. & Sommers, E. A. (Eds.), *Writing on-line: Using computers in the teaching of writing* (pp. 47–53). Upper Montclair, NJ: Boynton/Cook.

Marcus, S. (1987). Computers in thinking, writing, and literature. In Gerrard, L. (Ed.), *Writing at century's end* (pp. 131–140). New York: Random House.

McAllister, C., & Louth, R. (1987, March). *The effect of word processing on the revision of basic writers.* Paper presented at the annual meeting of the Conference on College Composition and Communication, Atlanta, GA. (ERIC Document Reproduction Service No. ED 281 232).

Moran, C. (1983). Word processing and the teaching of writing. *English Journal*, 72(3), 113–115.

Palmer, A., Dowd, T., & James, K. (1984). Changing teacher and student attitudes through word processing. *The Computing Teacher*, 11(7), 45–47.

Parson, G. (1985). *Hand in hand: The writing process and the microcomputer.* Juneau, AK: Alaska State Department of Education, Office of Instructional Services. (ERIC Document Reproduction Service No. ED 270 791).

Schanck, E. T. (1986). *Word processor vs. 'the pencil' effects on writing.* Unpublished master's thesis, Kean College of New Jersey. (ERIC Document Reproduction Service No. ED 270–791).

Selfe, C. L. (1985). The electronic pen: Computers and the composing process. In Collins, J. L. & Sommers, E. A. (Eds.), *Writing on-line: Using computers in the teaching of writing* (pp. 55–67). Upper Montclair, NJ: Boynton/Cook.

Sharples, M. (1985). *Cognition, computers and creative writing.* New York: Wiley & Sons.

Sommers, E. (1985). *The effects of word processing and writing instruction on the writing processes and products of college writers.* Unpublished manuscript. (ERIC Document Reproduction Service No. ED 269 762).

Teichman, M., & Poris, M. (1985). *Word processing in the classroom: Its effects on freshman writers.* Unpublished manuscript, Ploughkeepsie, NY: Marist College and IBM Corporation. (ERIC Document Reproduction Service No. ED 276 062).

Thomas, I. E. (1985, November). *Uses of the computers in teaching the composing process.* Annual Report of the NCTE Committee on Instructional Technology, presented at the annual meeting of the

National Council of Teachers of English, Philadelphia, PA. (ERIC Document Reproduction Service No. ED 265 571).

Watt, D. (1983). Word processors and writing. *Independent School*, 14, 41–43.

Womble, G. (1985). Revising and computing. In Collins, J. L. & Sommers, E. A. (Eds.), *Writing on-line: Using computers in the teaching of writing* (pp. 75–82). Upper Montclair, NJ: Boynton/Cook.

Suggested further reading

Brand, A. G. & Leckie, P. A. (1988). The emotions of professional writers. *Journal of Psychology*, 122, 5, 421–439.

Dalton, D. W. & Hannafin, M. J. (1987). The effects of word-processing on written composition. *Journal of Educational Research*, 80, 6, 338–342.

Dunn, B. & Reay, D. (1989). Word processing and the keyboard: comparative effects of transcription on achievement. *Journal of Educational Research*, 82, 4, 237–245.

Haas, C. (1989). Does the medium make a difference? Two studies of writing with pen and with computers. *Human Computer Interaction*, 4, 2, 149–169.

Hansen, W. J. & Haas, C. (1988). Reading and writing with computers: a framework for explaining differences in performance. *Communications of the A. C. M.*, 31, 9, 1080–1089.

Hovstaad, U. (1989). Computer assisted essay writing: an interdisciplinary project at Eikeli Grammar School. In Williams, N. & Holt, P. (Eds.) *Computers and Writing*. Oxford: Intellect Books.

Kellogg, R. T. & Mueller, S. (1990). A knowledge-based view of composing on a word-processor. Paper available from the authors, Dept. Psychology, University of Missouri-Rolla, Rolla, MO 65401, USA.

Peacock, M. & Breese, C. (1990). Pupils with portable writing machines. *Educational Review*, 42, 1, 41–56.

Shaver, J. P. (1990). Reliability and validity of measures of attitudes toward writing with the computer. *Written Communication*, 7, 375–392.

Williamson, M. M. & Pence, P. (1989). Word processing and student writers. In Britton, B. K. & Glynn, S. M. (Eds.) *Computer Writing Environments: Theory, Research and Design*. Hillsdale, NJ: Erlbaum.

Chapter 6

Teaching College Composition with Computers*

Stephen A. Bernhardt, Penny Edwards and Patti Wojahn

Many college English departments have begun using microcomputers in composition classrooms in the hope that they will improve student writing (especially revision processes), encourage better attitudes toward writing, and perhaps stimulate a collaborative learning environment. We have a wealth of self-report data suggesting that the introduction of computers does lead to a range of benefits for student writers (Arkin & Gallagher, 1984; Feldman, 1984; Hunter, 1983; Moore, 1985; Nash & Schwartz, 1985; Rodrigues, 1985; Sommers & Collins, 1984; Womble, 1985). But researchers have only recently begun to investigate through controlled studies these claims for the benefits of using computers in writing classes.

Researchers who have attempted to document the effects of computers on student writers have produced evidence that is at best inconclusive. Etchison (1986) found that computer classes made greater gains than control classes, with the computer students writing longer, better papers at the end of the term, even though they started out the term with scores well below those of the control group. Cohen (1986) reports that students who used word processing made 34% more revisions on end-of-term essays than did the control students, even though the computer students were writing with pen and paper (which he interprets as an argument for transfer of effect from computers to pen-and-paper writing).

In a program evaluation at Miami University of Ohio, Storms (1986) reports no qualitative improvement in the essays of computer students compared with control students, but he does report a range of perceived benefits, including favorable teacher and student reactions, out-of-class computer use by students in the experimental group, and more in-class writing in experimental sections. Hawisher (1987) reports no increase in revising activities for computer students; in fact, her pen-and-paper students actually revised somewhat more. She found no differences in quality between computer- and paper-generated essays by her college students. Cross and Curey (1984) report inconsistent findings on measures of attitude, performance, process strategies, and grades across computer and regular groups.

We are left with repeated observations that both students and teachers appreciate using computers in their writing classes and believe such use results in improved writing and revising, but the findings from controlled studies are

* *Reprinted from* Written Communication, *1989, 6, 108-133, © Sage Publications Ltd. Reprinted with kind permission of the authors and Sage Publications Ltd..*

at best equivocal. (For a more complete review of the research, see Bernhardt & Wojahn, in press.) Hawisher (1986) reviews the research and concludes that the inconsistency is a result of research designs that are either not well conceived or not comparable with other studies, making it difficult to generalize.

The present study is a program evaluation undertaken at Southern Illinois University—Carbondale to assess the broad, measurable effects of using computers to teach introductory college composition. A total of 24 classes were studied—12 control classes and 12 experimental—with the experimental computer classes meeting in the lab for half of their instructional time. Data on the success of the program were collected from a range of sources: pre- and posttests of student writing under both impromptu and take-home conditions; pre- and posttests of writing anxiety; records on attendance, tardiness, withdrawals, and homework and essay assignment completion; end-of-term course evaluation by both teachers and students; and self-report data collected from teacher meetings and teacher logs. (For discussion of writing program evaluation, see Davis, Scriven, & Thomas, 1981; Witte & Faigley, 1983.)

Methods

Physical setting

The study took place during fall 1986, the second year after construction of two new IBM PC computer labs at the university. The two adjacent labs each contained 32 PCs, with one lab set up as a classroom and the other as a drop-in lab. The computer classroom provided a PC for each student and software, including PC-Write (a word processing 'shareware' program, which each student copied for personal use), spelling and grammar checking programs, and a range of other general applications software. The lab (and the classroom when not scheduled for classes) was open long hours: from 8 a.m. until midnight during the first eight weeks of the term and from 8 a.m. until 2 a.m. during the last eight weeks of the term, as student demand on the PC labs increased. Throughout the term students had access to PCs, though during popular hours, in midafternoon and early evening, there was often a wait of up to an hour. The lab and classroom were staffed by student workers and full-time staff hired by Computing Affairs. Whenever possible, the English Department assigned a graduate assistant to the classroom to help with classes.

Subjects

In total, 24 sections of approximately 22 students per section were studied; all were first-semester, introductory composition classes in a two-semester sequence. We did not identify sections as computer classes in the class schedule in order to approach a random assignment of students to experimental and control groups. There were 12 teachers; each taught two sections, one computer and one regular. The schedule did not list teachers' names.

Each of the computer classes met in the computer classroom one day per week and in a regular classroom one day per week, with each meeting lasting 75 minutes. All classes met on a Tuesday/Thursday schedule, and, as often as possible, teachers taught both sections in either the morning or the afternoon.

Instructional setting

The computer and regular sections were in some ways the same course, having evolved within the same composition program. Instructors were generally expected to teach a process approach, with emphasis on revision through peer critiques and teacher commentary at various stages of writing. The recommended text was Scholes and Comley's *The Practice of Writing* (second edition); one teacher chose to use Moss and Moss's *The New Composition by Logic*. Through the composition training program and because of the general pedagogical orientation of the department, the teachers were encouraged to take a rhetorical approach to writing, stressing the importance of purpose and audience by having students experiment with a wide variety of writing tasks through a general course sequence: from personal, reflective writing through more objective, factual reporting, toward more critical, frequent informal writing in addition to six or seven longer, more formal texts. All teachers kept office hours and were encouraged to schedule individual conferences with students. The teachers often worked in the PC lab themselves and provided informal conferences to their students on demand.

As a group, the teachers in the study were experienced writing teachers who received generally strong evaluations, who displayed an interest in teaching with computers, and who were willing to be part of a study; five were lecturers and seven were experienced graduate student teaching assistants. Of the 12 teachers, seven had taught in the computer classroom during the previous year; these seven and an additional four others had attended a weekly seminar the previous fall on word processing and the teaching of writing; the twelfth teacher had taught composition with computers on another campus.

We did not urge teachers to attempt to teach the two courses in parallel fashion, with computers being the only variable. We recognized that the instructional setting would influence teaching strategies: the nature and timing of assignments, the use of the textbooks, the use of small groups, the frequency of in-class writing, and so on. At a preterm meeting, the teachers discussed what they anticipated might be different in a lab setting, with those who had taught in the lab offering advice to the others. We agreed to meet with the teachers throughout the term to discuss the sections, compare notes, solve problems, and share insights. For the first few weeks, we met every week, and then every other week during the latter part of the term.

Evaluative measures

Pre- and posttests of student writing. Samples of students' impromptu and revised writing were collected during the second and the fourteenth week of the 16-week course. For the impromptu sample, students were given one of the assigned tasks, told that the assignment would count as part of their course grade, and given 45 minutes to write in class. When they were finished, the impromptu drafts were collected and photocopied by the researchers, who then returned the photocopies with instructions to the students to take the drafts home over the weekend and revise them. Students were again told that the revision would count toward their course grades. Students had from their Thursday class meeting to the following Tuesday to revise their essays.

The posttest followed the same procedures. Computer students were given the impromptu task in the computer classroom and told they had the option of composing on disk. The drafts of all the students were either copied onto disk or photocopied. Students were again asked on Thursday to rewrite, using what they had learned in the course to improve their essays, and to turn in their revised work on Tuesday. Computer students were told they could type or work with pen and paper if they chose. Revisions were collected on Tuesday, either on paper or on disk.

Two writing tasks were used in a split-halves design, with half the classes writing on Topic 1 for their pretests, and half writing on Topic 2. Topics were then reversed on the posttest. The two writing tasks had been pilot tested during the previous term, with a third topic being discarded because some students had difficulty responding to the task. The two remaining topics were designed to be comparable: They were similarly worded; both called for analysis and use of data presented graphically; both were two-part questions, asking students to generalize from the data presented and also to suggest what other data might be useful in supporting further generalizations about each topic (see the Appendix for the two topics).

All full sets of two pretests and two posttests for each student were coded numerically for section, topic, student, and form. Complete sets included four samples from each of 146 computer students and 194 regular students. All essays were then keyed into disk files, with standard margins and spacing. Grammar, punctuation, paragraphing, and spelling were left as written. Words or passages of text that had been deleted on the paper copy were not keyed. If students indicated insertions or transpositions, these moves were carried out. To the extent possible, the texts were copied to disk without enhancing them but reflecting the intentions of the authors. All essays were then printed for a uniform appearance.

Seven readers were trained to rank sample essays, using papers from students in the study who had not completed all four tasks. A detailed scoring guide with criteria for each scale was developed and used throughout the scoring process. Two readers then independently scored each of the 1,360 essays using a combined holistic/analytic scale with four variables scored on a 1–6 scale and two variables scored on a 1–3 scale. Those variables scored from 1 (low) to 6 (high) were as follows:

- *holistic*: overall impression of quality
- *organization*: presence and effectiveness of discourse level planning, structure, signaling
- *support*: development of generalization with supporting detail, examples, evidence, reasoned positions
- *fluency*: use of various and clear sentences, effective subordination, coordination, and transition

Those scored from 1 (low) to 3 (high) were as follows:

- *conventions*: use of standard English syntax, spelling, punctuation
- *task*: degree of response to writing task as stated

Where scores differed by more than a single point on a scale, a third reader scored the essay. The tie-breaking third score was used with the original score closest to it. Scores for each essay were averaged, producing a single score on each measure in a 1–6 range for the first four categories and a 1–3 range on conventions and task.

Daly-Miller Writing Apprehension Test. We administered the Daly-Miller Writing Apprehension Test during the second week and again during the fourteenth week of class.

Records of withdrawals, attendance, tardiness, and assignment completion. We collected data on withdrawals from university registrar reports. A team researcher collected biweekly reports from each teacher on attendance, tardiness, and completion of major assignments.

Student evaluations. During the penultimate week of the course, students completed the short form of the Instrument for Reporting Course and Teacher Effectiveness in College Writing Classes, a 21-item questionnaire using a five-point scale to assess both course and teacher (Witte et al., 1981). Six additional questions were pilot tested during the previous spring term, revised, and then appended to the Witte et al. instrument. They concerned use of textbooks and handbook, getting to know classmates, writing grammatically correct papers, becoming a better writer, and enjoying writing. These additional questions were scored separately from those in the Witte et al. instrument, which underwent factor analysis during its development. Direct questions about computers were not part of either the Witte evaluation or our added items. No reference was made to computers.

An additional questionnaire was developed for the computer sections, with questions directly relevant to that experience. This questionnaire also used a five-point scale and was pilot tested during the previous spring term, with substantial revisions made for wording and usefulness of the questions. A brief set of open-ended questions allowed for further student comments. All questionnaires were administered by researchers who made it clear to students that teachers would receive no student feedback until after grades were filed.

Teacher evaluations. Ongoing records were compiled from notes from the biweekly teacher meetings and on an ad hoc basis as teachers commented to the researchers on their courses. Teachers were encouraged to keep logs throughout the term; four did so and these became a part of the data. At the end of the term, teachers completed a comparative evaluation of their two courses: 33 questions that had been developed through pilot testing the previous fall and spring. Additional written comments were also solicited.

Class observations. Both classes of each of two teachers were observed continuously throughout the term, with a detailed time line of minute-by-minute activities and an accompanying log of notes kept for each of the four sections. These data will be presented elsewhere as an observational study but are mentioned here because they did influence our sense of the program and discussions during teacher meetings.

Data collection. Students in both regular and computer sections were told during the first week of class that writing samples, course evaluations, and measures of attitudes about writing would be collected because the English Department was interested in learning how our programs were working and how they might be improved. It was stressed that our intention was not to evaluate individual students and that any data collected would be anonymous.

Research assistants, not the classroom teachers, administered all pre- and posttests, questionnaires, and end-of-term evaluations to ensure that all classes received the same instructions and were tested under the same conditions. Directions for testing and assessment were printed and read by the research assistants, who also offered standard answers to any questions.

Results

Pre- and posttests of student writing

Interrater reliabilities were calculated for the four 6-point scales using Pearson's r. Results were as follows: holistic, .74; organization, .72; support, .71; fluency, .66. With the two 3-point scales, conventions and task, percentage of exact agreement was calculated: conventions, 63%; task, 96%. The high percentage of exact agreement on task was the result of sharply defined criteria. The two-part task entailed generalizing from a data set and suggesting what other data would be necessary to form a complete position. Students responded to both parts of the task (3), to one but not the other (2), or did neither but responded in some way not called for by the task (1).

Table 6.1 presents regular and computer posttest means for impromptu, revision, and improvement scores (improvement = revision—impromptu). The

Table 6.1. Adjusted Posttest Means, Regular Versus Computer

		Impromptu	*Revision*	*Improvement*
Holistic	Regular	3.29	3.70	.41
	Computer	3.17	3.78	.61
Organization	Reg	3.31	3.65	.34
	Comp	3.19	3.68	.49
Support	Reg	3.31	3.70	.39
	Comp	3.26	3.84	.58
Fluency	Reg	3.27	3.70	.43
	Comp	3.29	3.71	.42
Conventions	Reg	2.24	2.35	.11
	Comp	2.26	2.38	.12
Task	Reg	2.47	2.50	.03
	Comp	2.37	2.52	.15

posttest scores were adjusted for pretest scores in order to eliminate the differences between the groups in their initial abilities. This allowed us to simulate an experiment where treatments are applied to groups of equal abilities. On most measures, students in the computer sections performed slightly worse than the regular students on the impromptu portion of the posttest. But the computer students then improved their scores on the take-home revision so that they ended up with revision scores somewhat higher than those of the regular students. In no case, however, is the difference between the groups on either impromptu or revision scores statistically significant.

The *improvement score* represents the difference between impromptu and revision, our measure of ability to improve a first draft through take-home revision. For the macro-level discourse features—holistic score, organization, and support—the computer students improved their essays to a greater extent than did the regular students. The two groups were comparable on their improvements of fluency and conventions; the computer group improved the task responsiveness of their essays to bring them to almost the same level as those of the regular students.

Table 6.2 presents the results of six separate analyses of covariance, where variance in improvement scores for each of the six scaled variables served as

Table 6.2. F Values for Analysis of Variance Using Improvement from Posttest Impromptu to Posttest Revision with Pretest Scores Controlled

	Method (DF=1)	*Instructor/ Method (DF=11)*	*Instructor (DF=11)*	*Pretest (DF=1)*
Holistic	4.59*	2.64**	3.71**	1.15
Organization	2.38	2.86**	4.10**	.16
Support	3.64	2.71**	3.79**	6.81**
Fluency	.00	1.15	1.11	.79
Conventions	.02	1.35	1.78	.27
Task	5.23*	1.92*	3.49**	4.23*

*$p < .05$; **$p < .005$.

the dependent variable. Three independent variables were defined in the model in addition to the error term; method (computer versus regular), instructor, and interaction or instructor with method. The covariate pretest score was included in the model to reduce variability of the experimental error. The degrees of freedom for all error terms was 305.

The instructor is the strongest effect, followed by the interaction of instructor with method, with four of the six scales showing significant contributions at the .005 level (df = 11). Even after accounting for the contributions of teacher and the interaction of teacher with method, method alone still contributes significantly to variance in posttest holistic improvement scores ($p < .05$; df = 1), and in task responsiveness ($p < .05$; df = 1), while the F values for organization and support approach significance. Improvement in fluency and conventions do not appear to be much affected by any of the factors in the model.

Note that the improvement scores on the pretest (students' abilities at the beginning of the term to revise and improve holistic scores, organization, and so on) do not appear to contribute significantly to the variance on posttest improvement scores, with the exceptions of support and task scales. Table 6.3 presents Pearson correlations for pretest improvement and posttest improvement scores. These low correlations suggest that ability to improve essays

Table 6.3. Pearson Correlations for Improvement Scores on Pre- and Posttests (n = 330)

Holistic	.07
Organization	.02
Support	.12
Fluency	.06
Conventions	.01
Task	.04

Table 6.4. t-Tests for Between Group Differences in Improvement Scores, Paper Versus Computer Revisers

	All Subjects (n=330)	*Paper Revisers (n=34)*	*Computer Revisers (n=112)*	*t*
Holistic	.48	.24	.69	-2.38*
Organization	.41	.15	.58	-2.22*
Support	.47	.22	.66	-2.31*
Fluency	.42	.26	.45	-.98
Conventions	.12	.00	.16	-1.36
Task	.09	.24	.12	1.18

**p*.05, assuming unequal variances.

through revising at the beginning of the term is not highly correlated with ability to do so at the end of term. In other words, learning appears to take place.

When the computer students wrote their posttest essays, they had the option of using either pen and paper or the computer to draft and revise their work. Table 6.4 presents t-tests of differences in improvement scores for paper revisers and computer revisers within the experimental treatment. The computer revisers improved their impromptu drafts during revision more than the paper revisers in every category except task. Those students who chose to revise on the computer at the end of the term significantly improved (at the .05 level of confidence) the overall holistic quality, the organization, and the level of support compared to those students who chose to revise on paper. The improvement scores of the computer revisers were larger, too, for fluency and conventions, though these measures did not achieve significance. The computer revisers' improvement scores were in every case larger than the average scores for the whole group of students in the study.

Daly-Miller Writing Apprehension Test

Results for the Daly-Miller Writing Apprehension Test are presented in Table 6.5. We used a response scale that reverses the normal orientation, so higher scores mean higher apprehension. Controlling for prescore differences and comparing least squares means yields an F value of 1.11 ($p > .29$; df = 1), a finding

Table 6.5. Test of Writing Apprehension

	(n)	*Pre*	*Post*
Computer Students	182	84.2	86.6
Regular Students	207	78.8	81.3

NOTE: Higher score - higher apprehension.

of no significant difference between computer and regular students on posttest anxiety scores. Both groups moved in the same direction at about the same magnitude, toward slightly increased anxiety.

Records of Withdrawals, Attendance, Tardiness, and Assignment Completion

Table 6.6 presents figures that compare regular and computer sections on withdrawals, attendance, tardiness, and assignment completion. Not all teachers kept complete records, so there are some missing data. In all cases, the data on the computer sections are somewhat worse than those for the regular sections. Though more regular students withdrew during the first week, the

Table 6.6. Withdrawals, Attendance, Tardiness, and Assignment Completion

Withdrawals	*Regular* (12 regular; 12 computer classes)	*Computer*
Total	33 (11.5%)	40 (14%)
First week	17	11
Weeks 2-5	5	12
Weeks 6-9	11	17
Attendance	*Regular* (10 regular; 10 computer classes)	*Computer*
Average % absent	9.02	9.51
First half term	8.15	8.62
Second half term	9.89	10.40
Tardiness	*Regular* (10 regular; 10 computer classes)	*Computer*
Average % tardy	3.12	4.39
First half term	2.36	3.78
Second half term	3.88	5.00

Percentage of students not turning in work:

	Regular	*Computer*
Major Assignments (11 regular; 11 computer classes)	4.18	8.45
Homework assignments (7 regular; 7 computer classes)	9.87	14.62

number of computer students withdrawing was higher during weeks 2–5 and 6–9, for a larger overall percentage. Attendance, too, was worse in computer sections, as was tardiness. Finally, the percentage of computer students not completing major writing assignments was double that of the regular students. Regular students also completed more of their homework assignments than did the computer students.

Student Evaluations

The short form of the Instrument for Reporting Course and Teacher Effectiveness in College Writing Classes loads all 21 items on two broad factors—teaching and course effectiveness. Table 6.7 presents the means for regular and computer groups. On average, the computer students rated their instructors slightly higher than did the regular students; the regular students responded slightly more favorably to questions concerning content. Neither difference was significant.

Evaluation of teaching. For the 12 matched sections, six instructors were

Table 6.7. Results of Course Evaluation

Teacher effectiveness	*Mean*
Computer	2.52
Regular	2.55
Course effectiveness	*Mean*
Computer	2.21
Regular	2.17

NOTE: These are grouped means for all items relating to teacher and all items relating to course effectiveness, calculated from a 1 to 5 scale, where 1 = 'strongly agree' and 5 = 'strongly disagree'. Polarity has been reversed on negatively worded items. Lower means indicate more positive evaluation.

favored by their computer sections, six by their regular sections. Of 18 statements concerning the instructor, ten were rated more positively by students in computer sections. The student responses most strongly favoring their computer teachers included items describing the teacher as (1) helpful, (2) good at using class time to help them as they wrote, (3) intellectually stimulating, (4) good at teaching how to support ideas with examples and details, (5) good at writing comments on paper that were easy to understand, and (6) a fair evaluator.

Students in regular sections responded more favorably to five of the 18 statements concerning their instructors. The student responses most strongly favoring their regular teachers included items describing the teacher as (1) good at trying to increase their confidence about writing, (2) good at teaching them to consider audience (responses were strong for both methods), and (3) good at teaching students to write different kinds of papers.

Evaluation of course content. A large majority of students in both computer and regular sections agreed that the course was currently useful to them, that what they learned in the course was valuable, and that the course would be useful to them in the future.

Students in regular sections appreciated their textbook and their grammar handbook more than students in computer sections, while more computer students agreed that they learned to write grammatically correct papers.

Students in computer and regular sections responded similarly to the statement that they got to know and work with their classmates. Comparable numbers also agreed that they were better writers at the end of the course than they were at the beginning. To the statement that they enjoyed writing more at the end of the semester, responses were almost evenly divided for both computer and regular sections.

Responses of computer students to the use of computers. An additional questionnaire asked computer students to evaluate the usefulness of computers for writing. Overwhelmingly, the students thought computers were a good writing tool. About half said that they preferred composing at the computer, with a third reporting that they did *not* usually write pen-and-paper drafts prior to using the computer. Two-thirds preferred to revise on the computer rather than with pen and paper. By the end of the semester, 78% agreed that they felt comfortable working with the word processor.

Responses showed that most students were not new to computers, with 43% saying they had used word processing prior to the course. Most students took advantage of their word processing skills for other courses, and a third reported teaching others to use PC-Write.

Supporting software was available to all students, yet the majority of students regularly chose to rely on the word processor alone. Not quite half used the spell checker often. Far fewer took advantage of the grammar software.

Many students (52%) felt that they spent more time on the course than their friends in regular sections. But about half also believed that they made better grades because of the word processor.

Although some people fear that the computer may dehumanize the writing classroom, the students in computer sections reported otherwise. Only 11% agreed that having computers in the classroom created a barrier between teachers and students. Only 16% felt that having computers in the classroom placed too much emphasis on machines and not enough on people.

Despite the distractions inherent in computer classroom settings, many students found the micro lab a comfortable place to write; a third disagreed. Nevertheless, two-thirds agreed that they liked having their class meet in the lab.

Responses of students to open-ended questions. Students' responses to open-ended questions were overwhelmingly positive. They liked the ease of revising, editing, and proofreading; the time they saved; the neatness of their papers at all stages; and the increased freedom in organizing. Most stated that using the computer caused them to change the way they planned, organized, wrote, revised, and edited their papers. They said that they now took more time to consider their writing seriously and more time to take their papers through various drafts. Many mentioned that spending extra time was worth it.

Access to the computers and lack of familiarity with function keys were the two problems or frustrations mentioned most frequently. Students overwhelmingly stated that lack of access to computers was the worst thing about writing with the computer. The second major problem was losing files. Students also complained that papers look different on screen than they do on paper, that it

Table 6.8. Responses of Teachers to the Use of Computers

	Mean	*Agree*	*Neutral*	*Disagree*
1. Students had problems getting to a machine:	4.67	12	0	0
2. I'd like to teach in the lab again:	4.58	10	2	0
3. I walked around the room more:	4.58	11	1	0
4. Students did all their assignments on the microcomputer:	4.42	5	7	0
5. I had to streamline and focus my lectures/ discussions:	4.42	11	1	0
6. The setting allowed for more interaction between me and the students:	4.25	11	0	1
7. The setting allowed for more interaction among students:	4.25	10	2	0
8. I required more in-class writing:	4.08	11	0	1
9. I made specific assignments that required the use of the microcomputer:	4.08	9	3	0
10. A lab assistant is necessary:	4.08	8	2	2
11. I enjoyed teaching this section more:	4.00	9	0	3
12. I lectured less:	4.00	8	3	1
13. Students revised more:	3.92	9	3	0
14. Students were more confident writing on the computer as opposed to pen and paper by the end of the course:	3.83	7	5	0
15. I checked the student writing more:	3.75	8	3	1
16. Students did more prewriting:	3.75	8	3	1
17. Students needed more instruction on the machine:	3.67	6	4	2
18. Student attitude was better:	3.58	8	2	2
19. I touched the students more:	3.50	6	5	1
20. I was stricter about the appearance of the papers:	3.50	6	4	2
21. I felt more in control of the class:	3.42	6	4	2
22. I felt closer to these students:	3.42	7	2	3
23. Grading was easier:	3.42	7	2	3
24. Students received less information about writing:	3.33	7	2	3
25. I required more prewriting:	3.25	5	4	3
26. I assigned more daily writing:	3.17	5	4	3
27. Students felt more anxious:	3.17	4	6	2
28. I felt more comfortable:	3.08	3	5	4
29. I created assignments on disk and transferred them to student disks:	3.08	6	0	6
30. I required more revising:	2.92	4	4	4
31. Student ideas were better:	2.75	0	10	2
32. Students turned in fewer late papers:	2.33	1	6	5
33. I required fewer major themes:	2.00	2	1	9

took those who didn't know how to type extra time to write papers, and that the lab was not the ideal place to concentrate on ideas.

The most typical advice students offered to their teachers or lab assistants was that they should better help students learn how to use the various commands and the available software. Other advice was to use the lab solely as a workshop environment, with no lecture time. Students also requested that not all minor assignments be done on computer, since access to computers was a problem.

Most felt the English Department should continue using computers, with some stating that all writing classes should meet in the lab. Students recognized the importance of computers in their futures: 83% said they would continue to use word processing.

Teacher evaluations

Responses of the 12 teachers to an end-of-term questionnaire are ranked in Table 6.8, beginning with items reflecting highest agreement and moving to lowest. To simplify presentation, 'strongly agree' and 'agree' are collapsed under 'agree'; disagreement categories are treated similarly. Possible means range from 5 (strongly agree) to 1 (strongly disagree). Most teachers indicated they preferred to teach in the computer classroom and all would like to do so again, feeling comfortable and in control of their classes. Most noted that student attitude was better and that the computer classroom was more conducive to helping students with their writing while it was still in progress. Teachers felt closer to their students and appreciated the workshop setting. The teachers felt strongly that a lab assistant was necessary and that more machine access would help.

In the computer sections, teachers required more in-class writing, slightly more prewriting, and more daily assignments. Teachers tended to agree that students were *doing* more revising, even though they were close to neutral on the question of whether they were *requiring* more revising. In the teachers' judgment, the students were comfortable using the word processor and actually did so for most assignments.

Discussion

In general, our results favor the use of computers in teaching composition. The pre- and posttest comparisons of impromptu writing show that the computer students did not perform quite as well as the regular students at the end of the term when asked to write an impromptu essay. But after revising, the scores of the computer group were higher than those of the regular group. The computer group revised their work to make significantly greater improvements during revision when compared to the control students. While significantly greater improvement was found for holistic scores, the computer students also improved their essays during revision (at levels approaching significance) on the high-level discourse features of organization and support. Significantly better task responsiveness was also characteristic of the revising of the computer students—during revision, they were more likely to bring their essays into accord with the assigned task. Very little effect was apparent for low-level features—sentence fluency and conventions. These findings suggest that the computer, considered by itself, does have a positive effect on student revising skills, especially discourse-level skills.

One reason the computer students were able to improve their essays more than the regular students was because their impromptu writing was generally poorer than that of the regular students. Perhaps becoming familiar with the machine's usefulness as a revising tool encouraged students to write quick first drafts, which they knew they could revise later. Whatever the explanation, the computer did help the students revise their work to the point where it was a little better than that of the regular students.

As a natural consequence of the experimental design, the computer students self-selected into two groups at the end of the term, those who chose to do their revising on screen and those who chose to revise on paper. Striking differences characterize these two groups' abilities to improve their essays on revision, with the computer revisers significantly outperforming the paper revisers on every measure except task responsiveness. All improvement scores for the computer revisers—including the holistic evaluation and the five analytic scales—were well above the average for all students in the study.

These findings suggest differences in adaptation to the technology, with some students (112 of 146, or 77%) becoming comfortable with the computer and finding ways to make it work as a revising tool. The important gains for this successful group would be obscured were their scores simply averaged with those of the smaller group who chose not to revise with the computer on the posttests. These findings echo Herrmann's (1987) findings with high school students; she identified three groups who adapted with varying success to writing on the computer. There does not seem to be a simple relation of machine to improvement; instead, one group takes ownership of the computer and uses it to good purpose, while a second group does not.

Though the data favored the use of computers considered in isolation, the covariate analysis showed stronger effects for teacher and for the interaction of the teacher with the method of using computers. The teacher had a very strong effect on whether the students improved. If our goal were simply improved student writing, we would probably get better results from choosing talented teachers or from training our teachers than from introducing computers into the classroom. If we do introduce computers, we need to work with the teachers to encourage adaptation to the technology. As other researchers have noted, we cannot simply put computers in a room and expect to see dramatic improvement in student writing and revising. Teachers, not machines, have the strongest effects on student writing improvement.

The attitudinal data suggest that both teachers and students viewed their experiences in the computer classroom positively. Few considered the computer a barrier between teachers and students, and few found the computer a hindrance to learning. Students seemed to appreciate learning both writing and word processing, skills they immediately applied both in and outside the composition classroom.

The teachers received slightly higher ratings from their computer sections than from their regular sections. This may reflect the teaching situation in a lab, where students receive more individual attention as they write—teachers can move about more freely, reading and discussing the student's writing as it is displayed on the screen. More positive student responses to instructors of computer sections may reflect the more intimate classroom interaction, with students and instructors working together more directly, allowing them to know and understand one another. Though the teachers felt the computer classes took longer to come together as learning groups, in the end, many instructors felt closer to their computer students. Teachers were certainly physically closer as they helped students write, sometimes offering a pat on the back, sometimes kneeling to make eye contact.

Teachers noted that students in their computer sections seemed to have better attitudes. Most believed that this was due to the ease of writing with a computer—students were able to locate problems in their writing and make changes quickly and neatly. Instructors also found students in computer sections more willing to revise and believed that the computer students did more prewriting and revising—even when it was not formally required. In spite of some negative feelings—about limited access to machines, about having to spend more time working on their essays, and about learning to use word processing (and invariably losing files)—the majority stated that they liked meeting in the computer classroom and that they liked using the computer to write.

There are certainly drawbacks to meeting in a lab environment. The students spend more time writing, but receive fewer lectures. Less lecture time in their computer classes caused instructors to feel rushed; some felt that the students in computer sections missed important material. However, the need to focus and organize lecture or discussion time might actually be viewed positively—many teachers reconsidered and streamlined their practiced lectures and usual assignments. And the quality of the students' writing did not appear to suffer from the shorter lecture time.

Further drawbacks are indicated by the consistently poorer showing of the computer students on measures of course withdrawal, attendance, tardiness, and assignment completion. From our pilot studies and from the reports of other researchers, we know that students enjoy meeting in the lab. We had often observed students arriving early and staying late, with some even staying through two full classes. So we were not prepared for data that showed that student attendance and tardiness were worse for the computer classes.

The instructors felt that the workshop atmosphere of the computer classroom affected student attitudes toward coming on time or, indeed, coming at all. While our teachers readily endorsed the value of a workshop classroom that focused on work in progress, students seemed slower to perceive the value. Many students preferred to write on their own time and viewed a workshop as 'nothing being taught'. Students also found they could easily enter the lab late and sit down at a computer, with little or no disturbance to others in the room.

Among those students who were in attendance, teachers repeatedly noted reluctance to engage in discussion or structured activities. In the lab, students expected to work on their texts with as little interference as possible. They did not want to discuss writing, do exercises, or answer questions on their reading. The computers exerted a strong draw on student attention, making it difficult for teachers to do anything except allow students to work individually.

The instructors noted in their logs and in weekly round-table discussions that their students often lost patience because of the long waiting period to gain access to computers. Some assignments may not have been completed because of limited time on the computers. Teachers and students agreed that free access to machines is important.

Some of the results of our study were surprising. We expected students in computer sections to see the class as more valuable and more useful to their futures, yet the responses to the evaluation questionnaire did not confirm this expectation. Likewise, we expected students in the computer sections to rank the class higher for its interactive aspect—working with other students—but such was not the perception. Again, students simply may not share our teacherly perceptions of the value of a collaborative class.

These results on student attitudes differ from our pilot study results, which indicated that students overwhelmingly preferred having their composition classes meet in a computer classroom. Pilot study students in computer sections had much more positive attitudes toward writing, toward the course, and toward their teachers than did the students in regular sections, giving both course and teacher more favorable ratings on every item but one. (The regular students in the pilot study found their handbook and textbook to be more useful than did the computer students.)

We believe the differences between the pilot study and the present study can be accounted for largely by the control of assigning each teacher to matched sections. Student evaluations appear overwhelmingly dependent on the teacher—the same teacher, even when methods are very different, receives approximately the same evaluation. This might suggest that we should view the glowing reports from other studies with caution—the more positive attitudes of students using computers in composition classrooms might largely be due to the teachers themselves, even when teachers in control sections are teaching from the same syllabus, using the same teaching approach, and so on. We might also assume a halo effect, with the initial enthusiasm among computer students and teachers giving way to more neutral attitudes after several terms.

While we think that matching teachers with sections successfully controlled for teacher effect, this control also exerted strong and unintended effects on the whole experience of computer classrooms. For the most part, instructors struggled to keep their computer sections parallel with their regular sections. Given a choice, most busy teacher/scholars prefer teaching one preparation twice over preparing for two different classes. If teachers had taught two computer sections, we believe we would have seen even more changes in strategies, assignments, and course requirements, and, we imagine, stronger effects on writing and attitudes. Instead, teachers worked to keep the students doing the same work on the same schedule. We saw a real change from the pilot study, when computer teachers talked about how they needed to reduce the number of assignments, change their strategies to fit the lab environment, and design assignments to take advantage of the collaborative possibilities of the computer classroom. During the study itself, we watched teachers do everything they could to make the classes the same, in spite of our urgings to adapt as necessary.

This study provides important support for using computers to teach college composition, and future studies should incorporate several features of this study. Students should be tested for revision skills as well as impromptu skills, since this is where we should expect to see the greatest benefits of the technology. Students also should be allowed to compose on computers for evaluation. While we should continue to recognize the importance of attitudes, we should also continue to collect data from writing evaluation and unobtrusive measures of course success (such as attendance or assignment completion).

Future studies should test systematically for the unintended results of this study. Specific studies should be designed to distinguish subgroups among students using computers to tease out the differences between those who adapt well to the technology and those who do not. Likewise, we should study and attempt to isolate what it is that determines how well individual teachers adapt to a lab environment, recognizing that we should not expect all teachers to be comfortable and successful in a lab setting. We should study how teachers change their strategies when they are free to adapt instruction to a lab setting by removing the constraint of double preparation. This suggests open, naturalistic investigation of teachers in lab settings to hypothesize and define variation and adaptation.

Finally, we need to acknowledge that we may not be able to measure confidently the effects of such a powerful writing tool on such a complex skill as writing ability. The literature on testing for student learning over the space of a term—whatever the experimental treatment and whatever skill is considered the dependent variable—is equivocal at best. The effects of computers on writing ability may not be a matter of quick transfer, but of subtle and incremental evolution over the life of a writer (see Perkins, 1985). The real results of introducing student writers to computers may be realized over the long term, as students continue to grow as writers and become increasingly proficient at using machines to enhance their writing processes and products.

Appendix: Topics for Pre- and Posttests

Topic 1

The World Almanac and Book of Facts 1985 (p. 113) presented the following information about various US cities:

Topic 1. Quality of Life in US Metropolitan Areas: A Comparative Table

City	*Per capita personal income 1982*	*% jobless Apr. 1984*	*Projected annual % growth in jobs 1979-1993*	*Projected annual growth in income 1979-1993*	*Mean # days clear-cloudy*	*Mean # days below 32 F*	*Normal daily Temp. Aug. F*
Atlanta, Ga.	11,590	4.3	3.1	3.5	108–149	58	86.4
Buffalo, NY	11,160	8.5	-	1.0	55–208	132	77.6
Chicago, Il.	13,069	8.7	0.6	1.6	86–173	99	82.3
Dallas-Ft. Worth, Tx.	13,846	3.9	2.7	3.6	138–132	41	96.1
Denver-Boulder, Co.	13,964	4.1	2.5	3.2	117–120	157	85.8
Honolulu, Ha.	12,130	4.7	1.8	2.4	86–100	0	87.4
Milwaukee, Wis.	12,597	6.6	0.7	1.6	94–172	144	79.7
Phoenix, Ariz.	11,086	3.9	2.9	3.6	213–071	10	102.2
Salt Lake City-Odgen Ut.	9,670	6.1	1.7	2.6	127–135	128	90.2
San Francisco, Ca.	17,131	6.2	1.4	2.3	162–104	0	68.2
Seattle-Everett, Wa.	13,239	8.4	1.5	3.0	71–201	16	74.0

The chart above summarizes information that could be used to determine which cities are the best places to live. Using the information given, discuss in an essay how these factors (such as per capita personal income) might relate to the quality of life. Also include suggestions concerning what additional factors are important in comparing the quality of various cities.

Topic 2

USA Today (May 5, 1986) asked American families about their vacation plans. They responded as follows:

Topic 2. American Family Vacation Plans

How much will you spend?	
500 or less	32%
501–1,000	25%
1,001–1,500	11%
1,501–2,500	13%
2,501-or more	12%
Don't know or wouldn't say	7%

*How will you travel?**	
airplane	53%
our car	50%
rental car	23%
train or bus	10%
camper or RV	8%

*more than one answer in some cases

What will you do?		Where will you go?	
Sightseeing	41%	Florida	13%
rest, relax	28%	California	10%
swim, water sports, sunbathe, beach	19%	Texas	5%
visit friends/relatives, socialize	14%	Bahamas	5%
play sports	10%	Europe	5%
fishing, hunting	10%	Canada	4%
camping/hiking	9%	Hawaii	3%
		Michigan	3%
		New York	3%

The chart above summarizes information about family vacations. Using the information given, discuss in an essay what seems to be the typical family vacation. Also include suggestions concerning what additional information is necessary to give a complete description of a typical family vacation.

References

Arkin, M., & Gallagher, B. (1984). Word processing and the basic writer. *Connecticut English Journal*, 15, 60–66.

Bernhardt, S., & Wojahn, P. (in press). Computers and writing instruction: A research review. In Jacobi, M. & Moran, M. (Eds.), *Research in basic writing: A bibliographic sourcebook*. Westport, CT: Greenwood.

Cohen, P. (1986, November). *Word-processing in freshman composition: A new study*. Paper presented at the annual meeting of the National Council of Teachers of English, San Antonio, TX.

Cross, J., & Curey, B. (1984). *The effect of word processing on writing*. (ERIC Document Reproduction Service No. ED 247 921).

Davis, B., Scriven, M., & Thomas, S. (1981). *The evaluation of composition instruction*. Inverness: Edgepress.

Etchison, C. (1986). A comparative study of the quality and syntax of compositions by first year college students using handwriting and word processing (Doctoral dissertation, Indiana University of Pennsylvania, 1985). *Dissertation Abstracts International*, 47, 163A.

Feldman, P. (1984). *Using microcomputers for college writing: What students say*. (ERIC Document Reproduction Service No. ED 244 298).

Hawisher, G. (1986). Studies in word processing. *Computers and Composition*, 4, 6–31.

Hawisher, G. (1987). The effects of word processing on the revision strategies of college students. *Research in the Teaching of English*, 21, 145–159.

Herrmann, A. (1987). An ethnographic study of a high school writing class using computers: Marginally, technically proficient, and productive learners. In Gerrard, L. (Ed.), *Writing at century's end: Essays on computer-assisted composition* (pp. 79–91). New York: Random House.

Hunter, L. (1983). Basic writers and the computer. *Focus: Teaching English Language Arts*, 9, 22–27.

Moore, W. (1985). Word processing in first-year comp. *Computers and Composition*, 3, 55–60.

Nash, J., & Schwartz, L. (1985). Making computers work in the writing class. *Educational Technology*, 25, 19–21.

Perkins, D. (1985, August/September). How information-processing technology shapes thinking. *Information Technology and Education*, pp. 11–17.

Rodrigues, D. (1985). Computers and basic writers. *College Composition and Communication*, 36, 336–339.

Sommers, E., & Collins, J. (1984). *What research tells us about composing and computers*. (ERIC Document Reproduction Service No. ED 249 497).

Storms, G. (1986, August). *Report on the Department of English computer classroom and laboratory 1985–1986*. College of Arts and Science, Miami University of Ohio.

Witte, S., et al. (1981). *The empirical development of an instrument for reporting course and teacher effectiveness in college writing classes*. Austin, TX: Writing Program Assessment Project.

Witte, S., & Faigley, L. (1983). *Evaluating college writing programs*. Carbondale: Southern Illinois University Press.

Womble, G. G. (1985). Revising and computing. In Collins, J. & Sommers, E. (Eds.), *Writing on-line: Using computers in the teaching of writing* (pp. 75–82). Upper Montclair, NJ: Boynton-Cook.

Suggested further reading

Anson, C. M. & Forsberg, L. L. (1990). Moving beyond the academic community: transitional stages in professional writing. *Written Communication*, 7, 2, 200–231.

Bernhardt, S. A., Wojahn, P. G. & Edwards, P. R. (1990). Teaching college composition with computers: a timed observation study. *Written Communication*, 7, 3, 342–374.

Gringrich, P. J. (1983). The Unix Writer's Workbench software: results of a field study. *Bell System Technical Journal*, 62, 6, 1909–1921.

MacDonald, N. H. (1983). The Unix Writer's Workbench software: rationale and design. *Bell System Technical Journal*, 62, 6, 1891–1908.

Nodine, B. (Ed.) (1990). Special Issue: Psychologists teach writing. *Teaching of Psychology*, 17, 1, 4–61.

Sterkel, K. S., Johnson, M. I. & Sjogren, D. D. (1986). Textual analysis with computers to improve the writing skills of Business Communication students. *Journal of Business Communication*, 23, 43–61.

Part III

Computers and special circumstances

Chapter 7

The Impact of Computers on the Writing Process of Learning Disabled Students*

Charles A. MacArthur

The computer is not a magical writing tool that will transform the way in which exceptional students write; neither is it a writing curriculum or an instructional method. However, it is a powerful and flexible writing tool with certain physical characteristics and information-processing capabilities that may affect the writing process and facilitate certain types of writing instruction. Computers can support the cognitive processes involved in planning, writing, and revising text. Equally important is the potential impact of the computer on the social context for writing in the classroom.

This article first discusses the key features of word processors and how they may affect the writing process and social context for writing. Next, a summary is presented of research evidence on the overall impact of word processors in schools. Finally, the article discusses the potential role in instruction of several extensions to word processors, such as spelling and style checkers, synthesized speech output, computer networks, and prompting programs that support planning and revising.

Features of word processing and their effects

Word processing differs from handwriting in several important ways that may influence the writing process. First, word processors permit flexible editing of text. Second, the visibility of the monitor and the use of a keyboard make writing more public. Third, they provide neat, printed copy. Fourth, they change the physical process of producing text, replacing handwriting with typing. Finally, word processors are complex tools that require some learning.The significance of each of these features is discussed in turn.

Flexible editing

The most often mentioned characteristic of word processors, or text editors, is the flexibility they provide in revising text. Changes in spelling, insertion and deletion of words and sentences, and large-scale movement of blocks of text can all be accomplished relatively easily. The potential impact of word proces-

---

* *Reprinted from* Exceptional Children, *1988, 54, 6, 536-542, © The Council for Exceptional Children. Paper reprinted with kind permission of the author and The Council for Exceptional Children.*

sors on revision is significant, since revision has been identified as both an important part of the composing process and a factor that distinguishes expert from novice writers. Though expert writers revise frequently to clarify meaning as well as to correct errors, the revisions of inexperienced writers are limited primarily to surface changes (Scardamalia & Bereiter, 1986). The ease of revision on the computer may encourage writers to make more revisions and improve their texts. It has also been suggested that the editing capability can affect the entire composing process by encouraging authors to write freely without concern for errors and awkward spots because it is so easy to make changes later (Daiute, 1985).

Some cautions are in order, however. The research evidence to date indicates that the impact of word processing on revision depends on individual writing skill. Revision is a complex cognitive process requiring writers to evaluate their writing, diagnose any problems, and figure out what changes to make (Flower, Hayes, Carey, Schriver, & Stratman, 1986). If students do not possess these cognitive skills, easing the physical requirements of revision will not help. Thus, it is not surprising that initial research indicates that experienced adult writers revise more extensively when using a word processor (Bridwell, Nancarrow, & Ross, 1984), but that word processing has limited impact on revision by inexperienced writers (Daiute, 1986; MacArthur & Graham, 1987).

Daiute (1986) reported that average eighth-grade students corrected more mechanical errors with a word processor than with pen and paper, but made few substantive changes within the text. MacArthur and Graham (in press), in a study of learning disabled (LD) students' composing, found no differences between handwriting and word processing in the overall number of revisions made by students, in the syntactic level of the revisions, or in the proportion of revisions that affected the meaning of the text. In both conditions, the majority of revisions were surface changes or minor changes in wording that did not affect meaning. The timing of revision, however, did differ between methods. With word processing, students made most of their revisions as they wrote the initial draft; whereas with handwriting, most revisions occurred when recopying the story. This difference suggests that, rather than freeing students from mechanical concerns during writing, the ease of editing may encourage writers to make many minor changes during initial composition.

Although word processing by itself appears to have little impact on revision by exceptional students, it may facilitate learning revising skills in an instructional context that teaches those skills. Graham and MacArthur (1988) taught LD students a strategy to use when revising opinion essays at a word processor. The strategy instruction increased both the overall number of revisions and the proportion of revisions that affected meaning, and also resulted in essays that were longer and higher in overall quality. Morocco and Neuman (1986) reported that a process approach to writing instruction combined with a word processor helped LD students learn to revise.

Visibility and social context

A second characteristic of word processors, less noted but perhaps equally important in instructional settings, is that the upright monitor and clear print make a student's writing accessible to peers and teacher and can promote social interaction around writing tasks. The accessibility of the monitor and the keyboard can be used to facilitate collaborative writing activities among students and sharing of work in progress (Levin, Riel, Rowe, & Boruta, 1985). Discussion of work with peers is a well-established principle of effective writing instruction (Graves, 1983). It should be noted that instruction on working cooperatively with peers is needed to ensure that collaborative writing activities are productive.

The visibility of writing on a word processor can also facilitate interaction between students and teachers (Morocco & Neuman, 1986). Teachers can observe the writing process of their students and gain a better understanding of how individual students approach writing tasks. Teachers can intervene at appropriate points to provide help with difficulties, to reinforce student decisions, or to react as a reader. Of course, the timing and content of teacher comments and questions are critical. Morocco and Neuman reported that special education teachers tended to intervene more actively when students wrote at a word processor, but that the impact on students depended on the teacher's approach to writing instruction. They found that students' motivation and sense of ownership of their writing were enhanced when teachers provided procedural support, or help in how to approach writing tasks, rather than giving substantive help with content or focusing prematurely on mechanics.

Printed copy

Word processors give students the power to produce neat, printed work and to correct errors without damaging the appearance of the paper. This aspect of word processing may be especially motivating for those exceptional students whose written work is typically characterized by poor handwriting and numerous mechanical errors.

Printed output may also encourage publication of work in a variety of formats for real audiences (MacArthur & Graham, 1987). Word processors and related software make it possible to produce letters, books, newsletters, and other publications with a professional look. Such publishing opportunities are valuable in establishing writing as a meaningful act of communication and in motivating student writing (Graves, 1983). When the teacher is the only audience, children may see writing as an exercise in correct form and display of knowledge—and as another opportunity for failure. When writing for a real audience, they start to see writing as a meaningful way of telling others about their experience and knowledge. Publication can also make all phases of the writing process more meaningful. For example, publishing a newspaper involves students in gathering and organizing information, selecting the most important points, writing clear descriptions, and revising and editing each others' work (Riel, 1985).

Typing

Typing is potentially an efficient way of producing text, especially for students with poor handwriting skills. Typing is not typically part of the elementary school curriculum, however, and most students find that typing is slower and requires more attention than handwriting. When typing is not automatic, it may interfere with higher order processes involved in composing and adversely affect students' writing. MacArthur and Graham (1987) found that typing proficiency was highly correlated with the length and quality of stories composed on a word processor. Our observations and those of others (Daiute, 1985) indicate that the slowness of typing can be frustrating for students and interfere with motivation.

Students need systematic typing instruction if they are to use word processors regularly. A reasonable goal, short of touch typing, is for students to use the correct fingering while looking at the keyboard and to achieve a rate at least equal to their handwriting. Brief instructional sessions can be included as a regular part of computer use. Several typing tutorials are available that provide carefully sequenced instruction, practice on phrases and sentences, and feedback on rate and errors. Teachers should monitor students to encourage them to use the correct fingering. Programs that emphasize games with time pressure should probably be avoided since they encourage students to abandon correct form for short term increases in speed.

Operation of a word processor

In addition to typing, students need to master the text-editing, filing, and printing operations of the word processor. The design of word processing software has improved in recent years both in power and ease of use, and several word processors have been designed specifically for use by younger students. Nonetheless, beginners of all ages commonly experience some frustrating difficulties in learning to use a word processor. MacArthur and Shneiderman (1986) described some of the problems that LD students have in mastering a word processor. One persistent problem area is misunderstanding the function of the return key in formatting text on the screen, which causes problems when students revise and print their work. Another common problem is loss of written work due to confusion about procedures for saving and loading files. Difficulties can be reduced by careful design of word processing software, selection of appropriate software for varying ages of students, and instruction in the operation of the word processor that anticipates common areas of difficulty (MacArthur & Shneiderman, 1986).

Overall impact of word processing

Motivation to write is often mentioned by teachers as a central reason for using word processors, and there seems little reason to doubt the numerous reports that word processing increases motivation (Daiute, 1987). In addition to improving motivation, two studies with LD students (Kerchner & Kistinger, 1984; Sitko & Crealock, 1986) reported that the use of word processing resulted in increases in the quantity and quality of student writing. Neither of these studies, however, compared the effects of special instruction in writing combined with a computer to special instruction without the computer, thus making it impossible to determine the contribution of the computer.

Research that has examined the effect of word processing independent of instruction has reported little impact on students' written products. MacArthur and Graham (1987) had fifth- and sixth-grade LD students, selected for their experience with word processors, write and revise stories using handwriting, word processing, and dictation. The handwritten and word processed stories did not differ on any of the product measures, including length, quality, story structure, and mechanical errors. Daiute (1986), in a study of nonhandicapped junior high students with extensive word-processing experience, found that the final drafts of word-processed compositions were somewhat longer than handwritten compositions and contained fewer mechanical errors but were not significantly different in overall quality.

Qualitative studies of the use of word processing in classroom settings indicate that the impact of computers on writing depends on the social and instructional context. Rubin and Bruce (1985) found that the effectiveness of word processing and related software depended on the decisions that teachers made about how to use the software and the social interactions that teachers permitted. In particular, they reported that the word processor facilitated social interactions among students if the teacher encouraged collaborative work. Morocco and Neuman (1986) found that word processors could be used to support a traditional skill-building approach to writing instruction, as well as an instructional approach focused on writing as a process. In the skill-building approach, word processors were used to present exercises and to correct mechanical errors in compositions. Within a process approach, word processors facilitated teacher-student interaction about the content of student writing and strategies for writing.

Research on word processing in school settings, especially with exceptional students, is still limited. Research is needed that examines the use of word processing with specific instructional techniques, such as instruction in revision, and with specific exceptional populations. Interactions among word processing, instructional methods, and the social context for writing also need further exploration.

Beyond word processing

The potential of the computer as a writing tool is not limited to word processing. Other computer applications, such as networks, spelling checkers and style analyzers, interactive prompting programs, and synthesized speech may also contribute to writing instruction for exceptional students.

Networks

Networks, both local area networks within a classroom and telecommunications networks, can offer expanded possibilities for written communication with real audiences. Peyton and Batson (1986) described the use of a network within a classroom to teach writing classes for hearing impaired students in which all discussion and interaction were conducted in writing. The network software enabled real-time conversation in writing. For hearing impaired students, the network provided an immersion approach to mastering English. The potential of the approach is not limited to hearing impaired learners. For hearing students, the approach can profoundly change the social context of writing and learning, facilitating collaborative writing and providing a connection between conversation and more formal writing.

Telecommunications networks can support written communication activities with distant audiences. Students need to write first for peers, parents, and teachers that they know in order to get direct feedback on how well their writing communicates (Graves, 1983). Students also need to write for less familiar audiences since a major way in which writing differs from conversation is that the audience is removed in time, space, and context (Scardamalia & Bereiter, 1986). The Computer Chronicles Newswire project (Riel, 1985) initially involved third and fourth graders with learning problems from three classes in Alaska and two classes in southern California and later expanded to include students from many countries. Students wrote articles about events and issues in their school and community and posted them on the network. Each site published a newspaper that consisted of articles selected from the network by the student editorial board. In the process, students entered into dialogues with others from different cultures, struggled with communicating clearly in writing, and gained valuable experience in evaluating and revising compositions. Cohen and Riel (1986) reported that essays written by seventh-grade students for other students via the newswire project were superior to essays written for the teacher to grade.

Spelling checkers and style analyzers

The analytical power of the computer can be tapped to help students with editing. Spelling checkers will check each word in a document and recommend possible spellings for any word not appearing in the program's dictionary. Sophisticated programs, for example, Writers Workbench (Frase, Kiefer, Smith & Fox, 1985), have been developed that will analyze aspects of style and grammar and provide editorial suggestions.

Spelling and style checkers have promise for exceptional students who typically have difficulty with spelling and mechanics, but further development of software designed for educational purposes, and of instructional methods, will be needed before computer analysis of writing will be helpful to beginning writers. Students can use spelling checkers to compensate for poor spelling skills, but current software was not designed to help students develop spelling skills. A spelling analysis tool designed for instructional purposes might look for common patterns in misspellings and provide that information to the teacher and student, or it might highlight only misspellings in a small set of words that an individual student is currently working on. Current style analysis programs were developed for business settings and are of limited usefulness for writers below the college level (Bridwell et al., 1984).

Interactive prompting programs

Several researchers have tapped the interactive capabilities of the computer to develop prompting programs to guide students in applying effective strategies for planning, writing, and revising. Most of the development work to date has addressed the prewriting stage, focusing on invention and organization. Burns and Culp (1980), for example, tested the effectiveness of a program that carries on a dialogue with college students to help them generate ideas on a topic. The program presents prompts based on rhetorical theory and has some limited capacity to respond to cues in the student's responses. The Quill writing system

(Rubin & Bruce, 1985) includes a Planner program that presents a series of questions designed to elicit ideas for an article. The prompts can be modified by the teacher for different types of writing. When used for a news article, for example, it might prompt students with who, what, where, and when questions. The student's responses are printed out for use in writing the news article.

Prompting programs have also been developed for use during composing and revising. Daiute (1986) used a word processor that included a revision prompting program. The program provided a set of questions that writers could ask themselves about the text they had just written, such as 'Does this paragraph make a clear point?' Based on the student response, the program offered general suggestions for improvement. Daiute (1986) compared students' writing on the word processor with and without the revision prompts and reported that the prompts led students to make more frequent and meaningful revisions; no data on overall quality were reported. Woodruff, Bereiter, and Scardamalia (1981) developed prompting programs to help students write opinion essays. Although middle school students liked using the programs and thought they were helpful, the programs had no effect on written products.

Scardamalia and Bereiter (1986) describe computer prompter programs as a form of procedural facilitation, aimed at easing the executive burden of writing by providing direct support in some aspect of the writing process.

Prompting programs could also be used within a strategy instruction approach to writing (Graham & Harris, 1987). Direct teacher instruction in a composing strategy could be followed by guided practice with a computer program that prompted students to follow the strategy.

Synthesized speech output

The first talking word processors were designed for visually impaired and vocally handicapped users, but recently word processors with synthesized speech output have been developed to support reading and writing activities for beginning readers and novice writers. Speech output permits inexperienced or poor writers to use their relatively stronger auditory language skills to monitor their written production. Rosegrant (1986) studied the use of a talking word processor with first, second, and third graders over a 6-month period. Students used the speech output to monitor the spelling of individual words as they wrote, to catch errors in syntax, and to listen repeatedly to their entire text. In comparison with students who used the word processor without speech, these students spent more time writing, made more revisions, and produced texts that were longer and higher in quality. Rosegrant theorized that hearing their writing helped students to develop a more 'critical ear', and thus to revise more effectively.

A talking word processor can support holistic approaches to reading and writing instruction that focus on meaningful communication rather than isolated skills instruction. Holistic methods must deal with the gap between what children want to express and what they have the skills to write and read. In initial language learning (the model for the holistic approach), adults support children in expressing themselves despite limited communication skills, but such individual scaffolding is difficult to provide in a classroom. A talking word processor can serve as a scaffold for both reading and writing, for example, by helping students read language experience stories and the writings of their peers.

Conclusions

Computers are dynamic tools for writing; they provide a wide range of opportunities for improving writing instruction. Word processors change the physical process of writing by replacing handwriting with typing and by making revision quick and convenient. Word processors and computer networks can change the social context for writing by supporting publishing for a variety of audiences and by facilitating collaborative writing projects and sharing of work in progress. Computers also can enhance instructional interactions between teacher and student by providing the teacher a window onto the writing processes of individual students. Interactive prompting programs can help students learn strategies for planning, writing, and revising. Synthesized speech can support reading and writing activities by exceptional students with limited reading skills. Spelling and style checkers can help students with the mechanical aspects of writing.

A caveat is in order. As with other educational applications of computers, the impact of computers on writing and writing instruction depends on how teachers and students make use of the technology. If computers are to contribute to better writing, they must be integrated with an effective instructional program. Special educators must develop sound instructional methods and computer-assisted composing tools that meet the needs of exceptional children. Further research is needed to determine how computers can be used most effectively to support writing instruction.

References

Bridwell, L. S., Nancarrow, P. R., & Ross, D. (1984). The writing process and the writing machine: Current research on word processors relevant to the teaching of composition. In Beach, R. B. & Bridwell, L. S. (Eds.), *New directions in composition research* (pp. 381–398). New York: Guilford Press.

Burns, H., & Culp, G. H. (1980). Stimulating invention in English composition through computer-assisted instruction. *Educational Technology*, 20(8), 5–10.

Cohen, M., & Riel, M. (1986). *Computer networks: Creating real audiences for students' writing (Report 15)*. La Jolla, CA: University of California, San Diego.

Daiute, C. A. (1985). *Writing and computers*. Reading, MA: Addison Wesley.

Daiute, C. A. (1986). Physical and cognitive factors in revising: Insights from studies with computers. *Research in the Teaching of English*, 20, 141–159.

Flower, L., Hayes, J. R., Carey, L., Schriver, J., & Stratman, J. (1986). Detection, diagnosis, and the strategies of revision. *College Composition and Communication*, 37, 16–55.

Frase, L., Kiefer, K., Smith, C., & Fox, M. (1985). Theory and practice in computer-aided composition. In Freedman, S. W. (Ed.), *The acquisition of written language: Response and revision* (pp. 195–210). Norwood, NJ: Ablex.

Graham, S., & Harris, K. (1987). Improving composition skills with self-instructional strategy training. *Topics in Language Disorders*, 7, 66–77.

Graham, S., & MacArthur, C. (1988). Improving learning disabled students' skills at revising essays produced on a word processor: Self-instructional strategy training. *Journal of Special Education*, 22, 131–152.

Graves, D. H. (1983). *Writing: Teachers and children at work*. Exeter, NH: Heinemann Educational Books.

Kerchner, L. B., & Kistinger, B. J. (1984). Language processing/word processing: Written expression, computes, and learning disabled students. *Learning Disability Quarterly*, 7, 329–335.

Levin, J., Riel, M., Rowe, R. & Boruta, M. (1985). Muktuk meets Jacuzzi: Computer networks and elementary school writers. In Freedman, S. W. (Ed.), *The acquisition of written language: Response and revision* (pp. 160–171). Norwood, NJ: Ablex.

MacArthur, C., & Graham, S. (1987). Learning disabled students' composing under three methods of text production: Handwriting, word processing and dictation. *Journal of Special Education*, 21, 22–42.

MacArthur, C. A. & Shneiderman, B. (1986). Learning disabled students' difficulties in learning to use a word processor: Implications for instruction and software evaluation. *Journal of Learning Disabilities*, 19, 248–253.

Morocco, C. C., & Neuman, S. B. (1986). Word processors and the acquisition of writing strategies. *Journal of Learning Disabilities*, 19, 243–247.

Peyton, J. K., & Batson, T. (1986). Computer networking: Making connections between speech and writing. *ERIC Clearinghouse on Language and Linguistics News Bulletin*, 10(1), 1, 5–7.

Riel, M. M. (1985). The computer chronicles newswire: A functional learning environment for acquiring literacy skills. *Journal of Educational Computing Research*, 1, 317–337.

Rosegrant, T. J. (1986, April). *It doesn't sound right: The role of speech output as a primary form of feedback for beginning text revision*. Paper presented at the annual meeting of the American Educational Research Association, San Francisco.

Rubin, A., & Bruce, B. (1985). *Learning with QUILL: Lessons for students, teachers, and software designers*, (Reading Report No. 60). Washington, DC: National Institute of Education.

Scardamalia, M., & Bereiter, C. (1986). Research on written composition. In Wittrock, M. C. (Ed.), *Handbook of Research on Teaching* (3rd ed., pp. 778–803). New York: Macmillan.

Sitko, M. C. & Crealock, C. M. (1986, June). *A longitudinal study of the efficacy of computer technology for improving the writing skills of mildly handicapped adolescents*. Paper presented at the Invitational Research Symposium on Special Education Technology, Washington, DC.

Woodruff, E., Bereiter, C., & Scardamalia, M. (1981). On the road to computer assisted compositions. *Journal of Educational Technology Systems*, 10, 133–148.

Suggested further reading

Exceptional Children, 54, 6, April 1988. Special Issue on Research and Instruction in Written Language.

Graham, S. & MacArthur, C. A. (1988). Improving learning disabled students' skills at revising essays produced on a word processor: self-instructional strategy training. *Journal of Special Education*, 22, 133–152.

Keefe, C. H. & Candler, A. C. (1989). LD students and word processors: questions and answers. *Learning Disabilities Focus*, 4, 2, 78–83.

Lynch, E. M. & Jones, S. D. (1989). Process and product: a review of the research on LD children's writing skills. *Learning Disabilities Quarterly*, 12, 2, 74–86.

Maarse, F. J., van de Veerdonk, J. L. A., van der Linden, M. E. A. & Pranger-Moll, W. (1991). Handwriting training: computer aided tools for remedial teaching. In Wann, J., Wing, A. M. & Sovik, N. (Eds.) *The Development of Graphic Skills*. London: Academic Press.

MacArthur, C. A., Hayes, J. A., Malouf, D. B. Harris, K. & Owings, M. (1990). Computer-assisted instruction with learning disabled students: achievement, engagement and other factors that influence achievement. *Journal of Educational Computing Research*, 6, 311–328

MacArthur, C. A., Schwartz, S. S. & Graham, S. (1991). A model for writing instruction: Integrating word processing and strategy instruction into a process approach to writing. *Learning Disabilities Research and Practice* (in press).

Morton, L. L., Lindsay, P. H. & Roche, W. M. (1989). A report on learning disabled children's use of word processing versus pencil and paper creative work. *Alberta Journal of Educational Research*, 35, 4, 283–291

Smith, A. K., Thurston, S., Light, J., Parnes, P. & O'Keefe B. (1989). The form and use of written communication produced by physically disabled individuals using computers. *AAC—Augmentative and Alternative Communication*, 5, 2, 115–124.

Vacc, N. N. (1987). Word processors versus handwriting: A comparative study of writing samples produced by mildly mentally handicapped students. *Exceptional Children*, 54, 2, 156–165.

Chapter 8

Computer-aided Writing and the Deaf

Transitions from Sign Language to Text via an Interactive Microcomputer System*

Keith E. Nelson, Philip M. Prinz and Deborah Dalke

In a typical school dealing with deaf children or other special education students, if computers are available for regular instruction they are microcomputers with random access memory (RAM) limited to 128K or less. For these circumstances there have been very few instructional software programs available that have truly interactive characteristics (Behrman, 1982; Rose & Waldron, 1983; Schwartz, 1984; Ward & Rostron, 1983). The present report concerns reading and writing instruction through a highly interactive program for Apple II computers with the stated RAM limitations. The program has been employed for five years with deaf children in a project termed the **alpha microcomputer-videodisc project for interactive reading, writing and communication**. A key component of this interactive project has been the provision to children of a special keyboard that displays words rather than letters. A child can easily and rapidly create new sentences and trigger lively graphics by touching the words that make up messages such as, 'the lion chases the elephant' or 'the car bumps the bicycle'. Another crucial and novel component in this project is the reliance on sign language as an aid to insuring that the text activities are sensible and productive for the child. Signs are provided on the TV screen through the software, and sign language exchanges occur between the teacher and the child as part of the interactive instructional flow.

In this ALPHA project, children receive instruction that is supplementary to their regular reading and writing activities. Positive impact on reading and writing and general communication can be expected from the following design characteristics of the microcomputer-interactive sessions:

1. Drill and practice is bypassed in favour of highly interactive activities in which children spend their time on their computer primarily in initiating messages that are displayed though text, pictorial animations, and sign animations on the TV screen. This sets a tone that prepares the way for side conversations between child and teacher that comment on, clarify,

---

* *Reprinted from the Seminar on Language Development and Sign Language held at Bristol University, 1986 and reprinted from Woll, B. (Ed.)* Language Development and Sign Language. *Monograph No. 1, International Sign Linguistics Association, Centre for Deaf Studies, University of Bristol, UK, 1989. Paper reproduced with kind permission of the authors and the International Sign Linguistics Association.*

and expand the messages from the TV so that they are meaningfully connected to knowledge and experience that the child has already accumulated (cf. 'recasting' in first language learning, Nelson 1980).

2. The heart of the child's learning occurs in a program mode called 'Create Sentences' in which the child tells the computer what to do by creating (writing) a text sentence. For example, if the child presses special word keys to write the sentence onto the TV, 'The cat chases the rabbit', the child can then in rapid order read the sentence on the TV, press one more key (a green 'go' key) and see the sentence acted out through pictorial animation. One more quick key press will then show the ASL signs that convey the same message as the newly-created sentence.

   This creation process allows the child to get a lot of satisfaction and motivation from the successful triggering of animations that make sense and that help the child to learn text. This sentence creation process also shows the child that producing complete sentences does not have to be a slow or boring process. In addition, the sequence described above takes advantage of special presentation capacities of the microcomputer that are usually absent in the child's encounters with text in books, handouts, exercises, notes, and so forth (cf. Lepper & Malone, 1986; Sheie, 1985).

   The microcomputer can display the child's own text message and within a matter of ten seconds or so present lively, colorful animations that show the meanings of the text. Seeing the meanings in a short time period in text, picture, and sign may aid the child in connecting these modes and in working out new understanding of the text material. At the same time, the special skills of the teacher are utilised because the teacher holds related side conversation with the child. These conversations may help the child to assimilate the text, sign, and pictures to knowledge the child has already acquired about cats, rabbits, chasing, and so on. Because teacher and child converse in the child's best communicative modes, the child is kept 'off the spot' and is free to use his or her current knowledge to best advantage in learning new text and in exchanging ideas with the teacher.

3. The computer helps greatly in assessment and record-keeping. The ALPHA programs have word tests that are used to find a lesson with a good level of challenge for each child at each stage of learning. In this present report as in past reports on other samples of children, it is typical for teachers to select lessons that include 30% to 70% correct word choices at the start of an ALPHA lesson. Enough exploration is then arranged by the teacher so that the child can be expected to have learned to deal flexibly with the text in the lesson. The child's explorations occur partly with individual word program modes, but as noted above, the primary exploration occurs with sentence creation. Following exploratory learning, the child is again assessed using a 'Sentence Test' mode in which pictures are animated on the screen and the child is given the demanding task of knowing the words as well as their appropriate organisation into a descriptive sentence—for example, 'The girl bumps the bicycle'. Additional assessments at the end of many lessons look at the child's ability to form such descriptive sentences from the word keyboard after viewing sign

language sentences on the TV screen. For both kinds of sentence tests, if the exploratory material has been used well the child can be expected to know most of the written words and to form descriptive sentences with accuracy exceeding 80%.

4. The use of signs in both child-teacher interaction and in language displays generated by the microcomputer provide the deaf child with a 'third channel' of meaning that is often taken for granted when hearing children read or generate sentences accompanied by graphics. The first two meaning channels are text and pictures, and the third meaning channel is the child's first and most fluent language (speech or sign, or both).
5. To deal successfully with written messages in books, magazines, and varied sources, the child must learn to deal with text as a 'formal code' (Olson & Nickerson, 1978; Cole & Scribner, 1974) that has a precise meaning. Sometimes that meaning will be fanciful rather than literal and realistic. As part of the ALPHA instruction, the children accordingly are expected to create and learn to understand both realistic and fanciful sentences.

The present report addresses three central questions concerning skill gains by a new sample of deaf children who used the ALPHA interactive system with their teachers during part of one school year:

1. Do the children read high mastery levels for text lessons presented within the ALPHA microcomputer-TV display modes?
2. Is there evidence of carry-over from reading gains on computer-presented lessons to gains on reading test batteries that are different from the computer instructional material? Although such carry-over may have occurred in prior years of ALPHA research, this report is the first to systematically assess reading battery performance prior to and following the interactive teacher-plus-child-plus-computer sessions.
3. Do children learn to read and write realistic-probable sentences as well as they read and write fanciful-improbable sentences? To the extent that this occurs, then the deaf students may have progressed beyond probabilistic strategies for text use toward solid control of text as a 'form' system in which meaning is indeed precise—a complete written sentence means exactly what it says, whether this meaning is familiar or unfamiliar, probable or improbable (cf. Olson, 1977; Olson & Nickerson, 1978).

Methodology

Subjects

The subjects in the current study included 63 deaf children between the ages of 3 years 4 months, and 11 years 8 months at pre-test (see Table 8.1). The subjects were enrolled in schools in a large metropolitan centre in the North Eastern region of the United States. All children had average or better intelligence as indicated by a performance scale of an accepted intelligence test. Most subjects have a severe to profound hearing impairment, but three subjects included have a moderate hearing impairment. An initial assessment was conducted of each child's basic communications skills, including sign language, fingerspelling and simultaneous communication (speech and sign). After the child's primary mode of communication was determined, reading and language measures were administered in that mode. These measures included a sentence imitation test with items that increased in length and grammatical complexity, a measure of grammatical comprehension of syntactic and semantic relationships, and a test of receptive and expressive vocabulary at the single word/sign level. Baseline measures of the child's reading abilities were also completed. Subjects were selected to participate in the study if they had beginning reading skills and at least preschool language levels, and if they could not already read the majority of the vocabulary in the ALPHA lesson set.

Table 8.1. Subject Profiles

Subject (n = 63)	*Age at Pretest (Years, Months)*	*Degree of Hearing Loss**
1	3.42	S/P
2	3.58	S
3	3.75	P
4	3.75	S/P
5	3.83	M/S
6	4.00	S/P
7	4.08	P
8	4.08	S/P
9	4.08	S/P
10	4.33	P
11	4.50	P
12	4.58	S/P
14	4.67	S
15	4.75	S/P
16	4.92	M/S
17	4.92	P
18	5.17	P
19	5.25	P
20	5.25	S/P
21	5.25	S/P
22	5.33	P
23	5.42	M/S
24	5.58	P
25	5.83	S
26	5.92	S
27	5.92	S/P
28	6.00	S/P
29	6.00	S/P
30	6.00	S/P
31	6.00	P
32	6.17	M/S
33	6.33	S/P
34	6.42	M/S
35	6.42	S/P
36	6.67	S/P
37	6.67	P
38	6.75	S
39	6.92	P
40	7.08	P
41	7.08	P

42	7.17	S/P
43	7.33	P
44	7.42	S/P
45	7.58	P
47	7.75	P
48	7.83	P
49	7.92	S/P
50	8.00	S/P
51	8.00	S/P
52	8.17	M
53	8.25	M
54	8.33	S/P
55	8.58	P
56	8.83	S/P
58	9.50	M
59	9.58	S/P
61	9.75	P
62	9.75	P
63	9.92	S/P
64	9.92	P
67	10.33	M/S
69	11.67	S/P
70	11.83	P

*M = Moderate, S = Severe, P = Profound

Equipment

Each subject worked with an Apple IIe microcomputer system (+64K bytes), which was set up with a special, large interface keyboard (Keyport 717) by Polytel Corporation, with interchangeable overlays that allowed more flexibility than a regular computer keyboard. Each overlay corresponds to one of the different lessons in Level I of the ALPHA Interactive Language Program. When the child selected word keys on the Keyport overlay the corresponding words or phrases were printed and displayed on a large colour monitor and appropriate graphics in pictures or signs also appeared on the monitor.

Software and instructional procedure

A software package—the ALPHA Interactive Language Series—was developed to assist deaf children and other special education children who are in the early grade levels of reading or who are just beginning to read. In the ALPHA program, the computer displays items in the form of printed words, animated pictures, and animated graphic representations of manual signs.Level I includes eight packages with eight individual lessons each. Lessons 1–6 introduce new vocabulary and Lessons 7 and 8 are review within each package.

The ALPHA courseware contains four primary modules: Individual Words; Creating Sentences; Testing Words; and Testing Sentences. Subjects were systematically exposed to the vocabulary of a given lesson within the Individual Words mode and then tested for knowledge of the words. The subject was then always exposed to the Creating Sentence mode where he/she could explore sentences by pressing keys on the interface keyboard and waiting briefly for the animated pictorial and sign sequences to appear on the TV screen. Finally, the teacher switched to the Testing Sentences format to assess the subject's level of competence in producing sentences.

Subjects were assigned to one of three microcomputer instructional groups—low, averaging 2.9 computer sessions; medium, averaging 25.0 sessions; and high, averaging 32.3 sessions and also more total sentences explored and tested. We used both number of sessions and total amount of work accomplished to distinguish the three computer experience groups, as this allowed us to compensate for the fact that children with long attention spans did more work per session than children with short attention spans. Additionally, subjects were divided into a younger group (3 years 4 months, to 6 years 4 months) and an older group (6 years 7 months, to 11 years 8 months). Each computer session averaged 15–20 minutes in duration.

During the course of instruction the set of words, pictures, and sign language graphics was gradually introduced. When the child pressed the appropriate word or sequence of words on the special keyboard, the computer responded accordingly. For example, if the child pressed the keys for 'The girl chases the chicken,' the video monitor would display both the text and redundant pictorial information. Then the same sentence could be represented with animated signs on the computer display. All sentences were active sentences with subject-verb-object structure. Conversational discourse was an integral part of the computer sessions as teachers were encouraged to request more information or comment on the child's message. This process could be implemented, for example, by telling the child something else about the chicken and displaying appropriate print and graphics. This procedure resulted in a dynamic interactive exchange of information with the child learning to communicate about the computer material and learning to gradually extend the new reading knowledge and vocabulary beyond the computer.

Reading and language assessment: pre-measures and post-measures

Prior to the introduction of computer-assisted instruction, all of the children participating in the study received a set of reading and language measures which were readministered during the final month of instruction. General language skills were assessed because we wanted to insure that the three groups compared—high and medium and low computer experience—were at highly similar levels in language as well as reading before instruction (pretest).

1. *Generalised Vocabulary Reading Test.* This measure consisted of a series of individual words and short sentences printed on flash cards which the child was required to read using his or her preferred language mode. Pretests were printed in one typeface and post-tests in another to assess the child's ability to transfer reading skills to a different printed form. Both pretest and post-test typefaces differed from the typefaces used for all the words appearing in computer instruction on the video display and on the children's special keyboard.
2. *Formal Reading Achievement Test.* A formal reading achievement test was administered to assess reading comprehension and reading vocabulary. Dependent on the age and language/reading levels of the child one of the following reading measures was administered:

The older higher functioning children in terms of language and reading skills were given two reading subtests from the Stanford Achievement Test (SAT) [Gardner, Rudman, Karlsen & Merwin, 1982]. The 'Word Reading' subtest, a measure of the child's ability to recognise words and attach meaning to them, was administered. The other subtest was the 'Reading Comprehension' measure, in which the child was asked to read short sentences increasing in length and grammatical complexity.

Children for whom the SAT was inappropriate were given the reading subtests from the Stanford Early School Achievement Test (SESAT) [Madden, Gardner & Collins, 1982]. Subtests administered included: (1) The 'Word Reading' subtest (involving a printed word matching task, matching a signed word to a word in print, and identifying a printed word representing a particular picture); (2) The 'Word and Stories' subtest which measures knowledge of word meanings; and (3) for relatively more advanced readers the 'Sentence Reading' subtest was given in which the child identified pictures which best illustrated the meanings of printed sentences.

For the youngest children and beginning readers an adapted version of The Test of Early Reading Ability (TERA) [Reid, Hresko & Hammill, 1981) was employed. The items included in this version of the TERA were: 'print awareness' in everyday situational contexts (child was required to read simple signs, logos, and words frequently encountered in the environment); 'relational vocabulary' (child was asked to select two words that are associated with a stimulus word); 'Discourse' (ie comprehension of a story which is signed to the child); and 'Alphabet Knowledge' which measures letter naming and recognition.

Language measures administered to all children were the following:

1. Sentence Imitation Test. This test measures initial responses in the child's best communicative mode (ie sign language or simultaneous communication). It includes items of increasing length and grammatical complexity. Sentences at low complexity levels include items such as 'Her coat is pretty'; and sentences at high levels include 'Bring me the car that is on the chair.'

2. Single Word/Sign Receptive and Productive Vocabulary Test. This measure was adapted from The Peabody Picture Vocabulary Test—Revised (PPVT) (Dunn & Dunn, 1981). It was used to assess both receptive and productive vocabulary at the single word and sign level.

3. Grammatic Comprehension Test. This measure which was adapted from The Assessment of Children's Language Comprehension (ACLC) (Foster, Gidden & Stark, 1972) assesses basic grammatical constructions and semantic relations.

Results

Performances with the microcomputer materials

The subject's progress in text encoding and decoding skills as well as general communication skills were formally evaluated from the data on the printers attached to the microcomputer, as well as from results of the specific reading and language measures. The printer provided highly reliable data in the form of printed records of all child key presses and teacher key presses. Additionally, at the termination of each instructional session the teacher completed a 'Daily Computer Record Form'. Teacher comments regarding the child's performance during computer sessions were either entered on the Daily Record Form or entered directly onto the printed record via the printer connected to the Apple microcomputer and the Keyport 717 keyboard.

Performance on the Computer-Generated Word Test is summarised in Table 8.2. The child's age and language level could be expected to influence test performance on the individual words because the test scores reported in this table came at the beginnings of lessons—before any exploratory learning. The results indicate significant effects for age on the computer-generated word tests: the older children (mean age of 8.17 years) performed significantly better than the younger children (mean age of 4.95). After mastering the individual words in a lesson, the child entered sentence exploration and was subsequently tested in this mode.

It was hypothesised that regardless of the chronological age and initial overall language levels of the subjects, all the children would perform well on tests assessing whether sentences were mastered. Our goal was to provide sufficient communicative interaction with the teacher and the computer to ensure mastery of sentences at a high level. As shown in Table 8.3, most children in both the younger and older groups performed at a high level on tests where picture and sign sequences occurred on the screen and the child 'wrote' a computer text answer. The older subjects performed at only a slightly higher level on tests where picture-sequences occurred on the screen and the child wrote a computer text response (mean of .85 for the younger group and .89 for the older group). Similar results were found when sign-sequences were displayed on the monitor and the child selected the corresponding printed words on the student keyboard (mean of .89 for the younger children and .91 for the older subjects). Thus, after word and sentence exploration the results demonstrated that younger and older subjects performed similarly on computer-generated tests.

Evidence on Carryover of ALPHA Gains to Generalised Reading Skills

To assess initial similarity between instructional groups, subjects received a language test battery including measures of grammatic comprehension, single word (sign) production and comprehension and sentence imitation. The researchers formulated a composite language score based on performance on the language measures. It showed that the high-, medium- and low-instruction groups were equivalent at pretest—with respective means of .75, .71, and .73.

Since it was hypothesized that overall reading and general language skills could be facilitated by the interactive computer approach to reading and

Table 8.2. Age group differences on computer-generated word tests before exploration (picture presented, child chooses word)

Group	*n*	*Group*	*n*
Younger Group[a] (3.4 to 6.4 years)	27	Older Group[a] (6.7 to 11.8 years)	23
Mean Percentage Correct 42%		Mean Percentage Correct 55%	

a. t-test for contrast between .42 and .55 $t = 3.14$, $p < .005$.
* Younger children had a mean age of 4.95 and a mean number of 30.3 computer sessions.
* Older children had a mean age of 8.17 and a mean number of 27.6 computer sessions.

Table 8.3. Performance after lesson exploration: similar scores by the two age groups computer-generated sentence tests

	Group Younger Group Mean Percentage Correct	*n*	*Group* Older Group Mean Percentage Correct	*n*
Computer Signs/ Child Chooses Text	89	25	91	19
Computer Pictures/ Child Chooses Text	85	26	89	23

* For each test mode all children with available scores in the younger group (mean age = 4.95 years) and older group (mean age = 8.17 years) were included. Age group performance differences were not significant.

writing instruction, we administered a reading and language test battery in addition to the computer-generated word and sentence measures. A composite reading score was determined based on performance on the reading measures described above. An analysis of variance was performed on the pretest-to-posttest patterns on the reading measures in relation to the amount of microcomputer instruction. Table 8.4 shows the relevant reading means as well as the average instructional time for the three instructional groups. On these reading measures there were differential gains by the instructional groups: low group, mean gain of 8.5; medium group, mean gain of 15.9; and high group, mean gain of 20.1. This expected interaction for pretest-to-posttest x instructional groups was statistically significant ($F = 3.99$, $p < .025$).

Table 8.4. Pretest scores and gain scores on reading measures in relation to amount of microcomputer instruction

Amount of Microcomputer Instruction *	*n*	*Reading Pretest*	*Reading Posttest*	*Reading Gain by Posttest*
Low	14	39.1	47.6	8.5
Medium	22	32.6	48.5	15.9
High	27	35.0	55.1	20.1

Anova for # Sessions Per Week X Pre-post. $F(2,60) = 3.99$; $p < .025$.
* Average number of computer sessions were 2.9 for Low Instruction, 25.0 for Medium Instruction, and 32.23 for High Instruction.

We also examined main effects of age on reading and language performance. Overall, the older subjects performed significantly better at both pretest and posttest on language and reading measures ($p < .0001$ in each case).

Evidence on fanciful and realistic sentences

Finally, we intensively analysed all computer records for a subsample of 16 children to determine if microcomputer tests in realistic and fanciful sentences yielded similar results. As anticipated, they did. Table 8.5 shows that the children performed nearly identically on realistic-probable and fanciful-improbable sentences.
In sum, reading and 'writing' (through key-pressing) skills were improved during the period of computer instruction for the deaf children in this study. Additionally, carryover gains in general reading ability were indicated by advances that increased as the amount and depth of computer instruction increased.

Table 8.5. Scores on computer-generated sentence tests by type of sentences

	Highly or moderately realistic sentence	*Highly or moderately fanciful sentence*
Computer shows pictures/ child chooses text	86.6	87.2

Discussion

The outcomes of the study provide positive answers to the three questions posed at the outset. First, it is clear that the deaf children did reach high levels of mastery on the ALPHA text material. For word tests and sentence tests presented on the computer-TV display the children averaged between 80% and 95% correct. These levels of achievement are in line with prior reports from previous samples of deaf children who participated in the ALPHA interactive language project (Prinz, Nelson & Stedt, 1982; Prinz & Nelson 1985a, 1985b; Prinz, Pemberton & Nelson, 1985).

A second outcome is based on the children's gains in reading outside the computer context. By pretesting and posttesting children on a general reading battery before and after the period of ALPHA instruction, it was possible to estimate generalisations or carryover from the interactive computer lessons to broader skill gains in the text area. Three groups of children were compared, all matched on initial levels of reading and language—high frequency of microcomputer instruction, medium frequency of microcomputer instruction, and minimal frequency of microcomputer instruction. Reading and writing instruction by the teachers was equivalent for the three groups when the supplemental ALPHA microcomputer lessons were not in use (the majority of the time for all groups). Generalised gains in reading were indicated by over twice as much average pretest-to-posttest advance for the medium and high ALPHA instructional groups as for the minimal comparison group. In short, deaf children who worked regularly on ALPHA interactive microcomputer lessons not only learned the ALPHA material but showed a carryover 'bonus' of stronger general reading gains than were shown by children who rarely or never worked with the ALPHA system.

The third outcome also represents a 'first' report for this project, part of a new phase of research in which clues about language processes will be sought in fine detail about children's performance. The finding that emerged is simply stated—children did equally well on computer-generated sentence tests regardless of the type of sentence examined, realistic-probable or fanciful-improbable. The overall rate of 87% correct on these sentences is impressive because the children must rely on their knowledge of English rather than on shortcuts or strategies or guesses from context to find the correct text. The skilled child in this testing context observes an action sequence on the TV and then proceeds to generate the appropriate text from the ALPHA keyboard, with equal facility for sentences such as 'The alligator chases the girl' and 'The girl eats the alligator'. It is clear from the test results that children are not confused by the necessity of dealing with the improbable sentences. Far from it. Instead children have an opportunity to talk with the teacher about reality-fantasy distinctions and to learn text as a formal code capable of expressing any meaning. We believe that there is a strong need in special opportunity to talk with the teacher about reality-fantasy distinctions and to learn text as a formal code capable of expressing any meaning. We believe that there is a strong need in special education for educational programs that help children to learn to use text in such flexible and powerful ways in the comprehension and production of both realistic and fanciful messages.

Results of the present study corroborate previous findings (Prinz & Nelson, 1985a, 1985b; Prinz, Pemberton & Nelson, 1985), that through ALPHA instruction deaf children as young as 3–5 years of age can acquire reading and writing skills. These findings fit well with previous claims that reading and writing need to be integrated within a system of exchange of information between teachers and children to be effective. Secondly, results regarding computer-generated 'writing' (through word key presses on the special student keyboard) support the notion that writing appears to complement and reinforce the acquisition of reading processes. Writing using the computer may serve as a catalyst for more active information processing and thus further facilitate reading processes.

As noted early in this chapter, rich discourse exchanges between the adults and the children were an essential part of the present microcomputer-assisted instruction, and these appear to have contributed to significant gains in processing written language. This finding is reflected in the pre to posttest gains on the text reading measures administered to the children. It further reinforces the notion that children naturally acquire the underlying principles of written language when they are exposed to text in a context of meaningful social interaction with others (Stokes & Branigan, 1984). By systematically embedding literacy training in the context of communicative interaction the childrens' motivation is greatly stimulated and they are able to more effectively make the connection between different forms of communication (ie signed and written). Teachers' elaborations and recasts that presented new language structures in reply to prior comments by the children could have aided the children's learning, just like these adult reply strategies aid first language learning (Nelson, 1977, 1980, 1987; Prinz & Masin, 1985).

The results also suggest that the adult-child sign language dialogues may have enhanced attention to and mastery of the text-picture and text-sign relationships on the video monitor. The children were highly motivated to encode and decode written messages since they could see that their own primary mode of communication (sign language) was represented in the materials presented on the computer. Furthermore, they could initiate innovative messages through sign language or text or a combination of both as well as respond to messages printed and/or signed by the teacher (cf. Lepper & Malone, 1986). The sign language graphics provided an appropriate, meaningful, and effective 'communication link' between the referent(s), the printed word(s), and the signed message(s).

The literature of the last 20 years shows that deaf children around the world commonly exhibit severe problems in learning how to read and write (Babbidge, 1965; Conrad, 1977; Bonvillian, Nelson & Charrow, 1976; Further, 1973; Gentile, 1972; Moores, 1982). One reason that deaf children continue to experience difficulty in learning to read and write may relate to traditional models of literacy instruction which are based on the correspondence between written forms (e.g. English or French) and speech. Many theorists have argued that phonological recoding converts the written words into a code based on their spoken forms (Menyuk 1976). It has been suggested that sensitivity to the orthographic structure of English and other Indo-European languages is developed through frequent exposure with the mapping between the orthography and the spoken language. Furthermore, it has been argued that the orthography is indifferent to dialectical or native language variation and that all readers are required to translate the orthography into their lexical-phonological representations to access word meaning (Menyuk, 1976). However, if a 'double' translation is required (that is, from a written text to a second language and then to the native language), then the task may not only be more difficult but will also depend on the accessibility of such translations to the reader (Chu-Chang, 1979). The ease with which second language learners develop and internalise text processing strategies may be highly dependent on the degree of familiarity they have with the lexicon of the second language. The text-learning situation relative to the deaf child appears to be similar in many ways to that of a hearing individual acquiring a second language (Ewoldt, 1981).

Recently researchers have begun to investigate alternatives to phonological recoding during reading for deaf people who rely primarily on sign language input. Central to the current study is an important line of psycholinguistic research that has recently emerged in the literature and which concerns the internal coding of text as associated with different aspects of language processing (e.g. lexical access, short-term memory and spelling). Research in this area has begun to reveal some underlying processes for text learning which appear to be mediated in deaf children and adults by sign and fingerspelling codes. Research with American Sign language (ASL) has indicated that this manual language can be internalised and used as a short-term memory code. Short-term memory studies have indicated that when ASL handshapes and signs are presented to adults they are retained in a sign-based code (Bellugi, Klima & Siple, 1975; Hanson, 1982; Locke & Locke, 1971; Poizner, Bellugi & Tweney, 1982; Shand, 1982).

The question of whether a sign language code can facilitate the acquisition of reading and writing processes has not been systematically investigated. A few studies have demonstrated that some deaf children are able to read and write 'manually' prior to having formal written language instruction (Bellugi, Klima & Siple, 1975; Hirsch-Pasek, 1981; Maxwell, 1983). These researchers have shown that the children are not taught literacy skills in the formal sense; rather, they learn text encoding and decoding skills through a combination of fingerspelling words and using signs in relation to written sentences. In a recent study, Hanson, Liberman and Shankweiler (1985), (as reported in Hanson, 1985), examined short-term memory coding in relation to beginning reading strategies in deaf college students. The researchers contrasted reading performance by good and poor readers and found that the good readers utilised linguistic codes that were both speech-based and manually-based. Furthermore, in an investigation of the role of sensory attributes in a vocabulary learning task in American Sign Language by deaf and hearing adults, Siple, Caccamise and Brewer (1982) found that the linguistic structure of signs facilitates the acquisition of new lexical items.

A related line of research concerns the sign coding of written words. Research here suggests that lists of words which have close sign language correspondences are easier to learn than those which do not have these close sign correspondences (Odom, Blanton & McIntyre, 1970). Similarly, Conlin and Paivio (1975), in a paired association task, and Bonvillian (cited in Bonvillian, Nelson and Charrow, 1976), in a free recall task found that deaf subjects learned easily signable items more readily than items for which there were no close sign correspondences. Deaf children who do not have early exposure to either fingerspelling or sign language appear to have generally lower reading skills compared to children with such exposure (Meadow, 1968; Vernon & Koh, 1970). Overall, then, the preliminary research suggests the need to more effectively integrate sign language and fingerspelling into reading and language instruction.

During the past five years, we have been conducting research on sign language, the development of text skills, and the use of multiple communication codes by children. A central and innovative component of this research has been the use of the ALPHA microcomputer technology in presenting a combination of text, sign language, and animated pictorial graphics, in appropriate contexts to facilitate communication. The research has involved the development of text encoding and decoding skills and general communication in deaf children using this ALPHA Interactive Language Series. The object of the research was to use supplementary teacher-plus-computer instructional sessions to achieve gains in general communication skills and in reading and writing that will serve the individual well regardless of his/her primary mode of communication. Results of earlier research have shown a significant improvement by deaf children in word and phrase identification, reading comprehension, and basic sentence construction or writing (Prinz & Nelson, 1985a; Prinz, Nelson & Stedt, 1982; Prinz, Pemberton & Nelson, 1985). Additionally, when tested on general communication, the deaf children who used ALPHA regularly also made skill gains in this area (Prinz & Nelson, 1985; Prinz, Pemberton & Nelson, 1985).

The goal of the present research was to use supplementary teacher-plus-computer computer-assisted instructional sessions to develop higher level reading and writing skills in severely to profoundly deaf children. The approach incorporated the use of a complete and novel interactive computer system that allowed the child to initiate communication to a teacher about a topic of interest by means of a combination of text on a special interface keyboard, text on the video monitor, and any already available language modes (sign language, fingerspelling). It appears that this supplementary instruction helped to fill the gap for these children in text skills, but much more remains to be done.

Implications for future research

If a deaf child is going to read or write fluently, it is essential to have text and related codes that are not just learned but represented with high accessibility. In order to study this, researchers need to go beyond accuracy of text reading and text learning to measures of flexibility. With deaf children there is virtually no such data available. What is needed is evidence on how well consolidated and how highly accessible text skills are in deaf children and, for comparison, in hearing children. The researchers are in the process of developing a series of experiments designed to test for the possibility of sign or 'cherological' (Stokoe, 1960) coding in severely to profoundly deaf children who rely primarily on sign language for communication when presented with [1] lists of individually printed words; [2] animated pictures and [3] signs from their native sign language presented on a microcomputer. The expectation is that sign coding will occur (cf. Hamilton, 1985; Shand, 1982) and that individual differences in efficiency of coding across stimulus modes will relate to patterns of skill in reading and writing.

It is noted in the literature on deafness and communication that most research is directed to a specific area or question and does not effectively deal with the interactions in spoken, written and signed communication (Prinz, 1985). It has become evident that a vehicle is needed which will permit investigation of a variety of linguistic and communicative codes and their interactions occurring in different languages. Microcomputer technology is one very effective way to rapidly display information (text as well as graphic representations of pictures and signs), record and time responses, and then critically analyse that response data.

In the teaching of reading and writing it can be expected that techniques very similar to recasting for spoken (Nelson, 1977, 1980, 1986) and sign languages (Prinz & Masin, 1985) can be used as facilitators for next text skill acquisition. Using the computer text display, pencil and paper, or other text display techniques, an adult can encourage the child to write a sentence (however imperfectly) and can then respond to the sentence by a written sentence recast very similar in meaning but with structural information that goes beyond the child's current written language system. Naturally the child and teacher can also use their first languages as part of this ongoing interaction. As of yet we have seen no tests of this hypothesis that written language recasts can be powerful facilitators of growth of written language structures. The researchers are developing tests of the hypotheses for deaf children whose first language skills are predominantly in sign language. In addition researchers need to consider cases in which deaf children or adults become multilingual by learning to map their first language skills in sign language not only to one written language but also to one or more 'foreign' written languages (e.g. a deaf individual in Sweden or Germany may become fluent in reading English and French).

Finally, the present researchers recently have expanded the ALPHA program by incorporating interactive videodisk technology (Nugent & Stone, 1982). In addition to text and picture animations, the child also sees real life pictures and sign language (ASL) from the videodisk. We plan to contrast traditional approaches of teaching written language to the use of such instruction combined with supplemental computer-videodisk system which incorporate signs from other sign languages (e.g. German Sign Language, British Sign Language, French Sign Language). It is anticipated that elaborated versions of this microcomputerised videodisk system could have wide application in educational programs for deaf students.

Acknowledgements

This research was supported by two Technology Effectiveness Research Grant (Numbers G008302959 and G008430079) from the US Department of Education (Office of Special Education and Rehabilitative Services) and a gift from the Hasbro Children's Foundation. Many individuals (including computer programmers, graphic artists and teachers) contributed to this study. The authors would like to mention Vic Broderick, Bob Coleman, Edward Henry, Michael Knight, and Reza Samiei for their assistance in programming the lessons. We thank Cathy Fuller for her skills in organising material and in drawing pictures and graphic representations of signs. We would like to thank Susan Price, Rose Angela, and Gary van der Meer, project teachers who participated in the field testing of the programs reported here. We are also grateful to Elisabeth Ann Prinz for assistance in selecting and correcting signs for the microcomputer lessons. The current field coordinators, Virginia McNamara and Ann Landy, also assisted in many ways. Finally, we would like to express our thanks to the children and teachers for their cooperation in implementing the ALPHA project.

References

Arcanin, J. and Kawolkow, G. (1980), 'Microcomputers in the service of students and teachers—computer assisted instruction at the California School for the Deaf: An update' *American Annals of the Deaf*, 125, 807–813.

Babbidge, H. (1965), *Education of the deaf in the United States*. Report of the Advisory Committee on the Deaf. Washington, DC: Government Printing Office.

Baker, N. D. and Nelson, K. E. (1984), 'Recasting and related conversational techniques for triggering syntactic advances by young children', *First Language*, 5, 3–22.

Bates, M. and Wilson, K. (1980), Language Instruction without Prestored Examples. Paper presented at the Third Canadian Symposium on Instructional Technology, Vancouver, BC, Canada.

Behrman, M. (ed.) (1984), *Handbook of microcomputers in special education*. San Diego, CA: College-Hill Press.

Bellugi, U., Klima, E. and Siple, P. (1975), Remembering in Signs. *Cognition*, 3, 93–125.

Bonvillian, J. D., Nelson, K. E., and Charrow, V. R. (1976), Language and language-related skills in deaf and hearing children. *Sign Language Studies*, 12, 211–250.

Chu-Chang, M. (1979), The dependency relation between oral language and reading in bilingual children. Unpublished doctoral dissertation, Boston University.

Cole, M. and Scribner, S. (1974), *Culture and thought*. New York: John Wiley & Sons.

Conlin, D. and Paivio, A. (1975), The associative learning of the deaf: The effects of word imagery and signability, *Memory and cognition*, 3, 335–340.

Conrad, R. (1977), The reading ability of deaf school learners. *British Journal of Educational Psychology*, 47, 138–148.

Dunn, L. and Dunn, L. (1981), *Peabody Picture Vocabulary Test—Revised (PPVT)*. Circle Pines, MN: American Guidance Service.

Ewoldt, C. (1981), Factors which enable deaf readers to get meaning from print. In: Hudelson, S. (ed.) *Learning to read in different languages*. Washington, DC: Centre for Applied Linguistics.

Foster, Gidden and Stark (1972), *Assessment of children's language comprehension*. Palo Alto, CA: Consulting Psychologists Press.

Furth, H. (1973), *Deafness and learning: A psychological approach*. Belmont, CA: Wadsworth.

Gardner, E. F., Rudman, H. C., Karlsen, B. and Merwin, J. C. (1982) *Stanford Achievement Test*. San Antonio, TX: The Psychological Corporation.

Gentile, A. (1972), Academic achievement test results of a national testing program for hearing impaired students, United States: Spring 1971 Washington DC: Annual Survey of Hearing Impaired Children and Youth, Office of Demographic Studies, Gallaudet College, Ser. D., No. 9.

Hamilton, H. (1985) Linguistic encoding and adult-children communication, In: Martin, D. S. (Ed.) *Cognition, education, and deafness*. Washington, DC: Gallaudet College Press.

Hanson, V. L. (1985), Cognitive processes in reading: where deaf readers succeed and where they have difficulty. In Martin, D. S. (Ed.) *Cognition, education and deafness: Directions for research and instruction* (pp. 108–110). Washington, DC: Gallaudet College Press.

Hanson, V. L. (1982), The use of orthographic structure by deaf adults: Recognition of fingerspelled words. *Applied Psycholinguistics*, 3, 343–356.

Hirsch-Pasek, K. A. (1981), Phonics without sound: Reading acquisition in the congenitally deaf, Doctoral dissertation, University of Pennsylvania.

Lepper, M. R. and Malone, T. W. (1986), Intrinsic motivation and instructional effectiveness in computer-based education. In: Snow, R. E. and Farr, M. J. (Eds.) *Aptitude, learning, and instruction, Vol III*. Hillsdale, MJ: Lawrence Erlbaum Associates.

Locke, J. L. and Locke, V. W. (1971) Deaf children's phonetic, visual and dactylic coding in a grapheme recall task. *Journal of Experimental Psychology*, 89, 142–146.

Madden, R., Gardner, E. F. and Collins, C. S. (1982), *Stanford Early School Achievement Test*. San Antonio, TX: The Psychological Corporation.

Maxwell, M. (1983), Language acquisition in a deaf child of deaf parents: Speech, sign variations and print variations. In: Nelson, K. E. (Ed.) *Children's language, Vol 4*. Hillsdale, NJ: Lawrence Erlbaum Associates.

Meadow, K. P. (1968), Early manual communication in relationship to the deaf child's intellectual, social, and communicative functioning. *American Annals of the Deaf*, 113, 29–41.

Menyuk, P. (1976), Relations between acquisition of phonology and reading. In: Guthrie, J. (Ed.) *Aspects of reading* (pp. 89–111). Baltimore, MD: Johns Hopkins University Press.

Moores, D. F. (1982) *Educating the deaf: Psychology, principles, and practices* (2nd edition). Boston, MA: Houghton Mifflin.

Nelson, K. E. (1977), Facilitating children's syntax acquisition. *Developmental Psychology*, 13, 101–107.

Nelson, K. E. (1980), Theories of the child's acquisition of syntax: A look at rare events and at necessary, catalytic, and irrelevant components of mother-child conversation. *Annals of the New York Academy of Sciences*, 345, 45–67.

Nelson, K. E. (1987) Some observations from the perspective of the rare event cognitive comparison theory of language acquisition. In: Nelson, K.E. and van Kleeck, A. (Eds.) *Children's Language, Volume 6*. Hillsdale, NJ: Lawrence Erlbaum Associates.

Nugent, G. C. and Stone, C. G. (1982), The videodisc meets the microcomputer, *American Annals of the Deaf*, 127, 569–572.

Odom, P. B., Blanton, R. L. and McIntyre, C. K. (1970) Coding medium and word recall by deaf and hearing subjects. *Journal of Speech and Hearing Research*, 13, 54–58.

Olson, D. R. (1977), From utterance to text: The bias of language in speech and writing. *Harvard Educational Review*, 47, 257–281.

Olson, D. R. and Nickerson, N. (1978), Language development through the school years: Learning to confine interpretation to the information in the text. In: Nelson, K. E. (Ed.) *Children's language, Volume 1*. New York: Gardner Press/Halsted.

Poizner, H., Bellugi, U. and Tweney, R. (1981), Processing of formational semantic information. *Journal of Experimental Psychology*, 7, 1146–1159.

Prinz, P. M. (1985), Language and communication development, assessment and intervention in hearing-impaired individuals. In: Katz, J. (Ed.) *Handbook of clinical audiology* (3rd edition). Baltimore, MD: Williams and Wilkins.

Prinz, P. M. and Masin, L. (1985), Lending a helping hand: Linguistic input and sign language acquisition in deaf children. *Applied Psycholinguistics* (in press).

Prinz, P. M. and Nelson, K. E. (1985), A child-computer-teacher interactive method for teaching reading to young deaf children. In: Martin, D. S. (Ed.) *Cognition, education and deafness*. Washington, DC: Gallaudet College Press.

Prinz, P. M., Nelson, K. E. and Stedt, J. (1982), Early reading in young deaf children using microcomputer technology, *American Annals of the Deaf*, 127, 529–535.

Prinz, P. M., Pemberton, E. and Nelson, K. E. (1985), ALPHA interactive microcomputer system for teaching reading, writing and communication skills to hearing-impaired children. *American Annals of the Deaf*, 130(5), 444–461.

Reid, D. K., Hresko, W. P. and Hammill, D. D. (1981), *The test of early reading ability*. Austin, TX: Pro Ed.

Rose, S. and Waldron, M. (1983), Microcomputer use in programs for hearing-impaired children: A national survey. *American Annals of the Deaf*, 129, 338–342.

Schwartz, A. (Ed.) (1984), *Handbook of microcomputer applications in communication disorders*. San Diego, CA: College-Hill Press.

Shand, M. (1982), Sign-based short-term coding of American Sign Language signs and printed English words by congenitally deaf signers. *Cognitive Psychology*, 14, 1–12.

Sheie, T. (1985), Integration and implementation: A four-point mainstream model. *American Annals of the Deaf*, 130, 397–401.

Siple, P., Caccamise, F. and Brewer, L. (1982), Signs as pictures and signs as words: Effect of language knowledge on memory for new vocabulary. *Journal of Experimental Psychology: Learning Memory and Cognition*, 8, 619–625.

Stokes, W. T. and Branigan, G. (1984), Operating principles in the acquisition of literacy. Paper presented at the Third International Congress for the Study of Child Language, Austin, Texas.

Stokoe, W. C. (1960), Sign language structure: An outline of the visual communication systems of the American deaf. *Studies in Linguistics, Occasion Papers 8*.

Vernon, M. and Koh, S. (1970), Effects of early manual communication on achievement of deaf children. *American Annals of the Deaf*, 115, 527–536.

Ward, R. D. and Rostron, A. B. (1983), Computer assisted learning for the hearing-impaired: An interactive written language environment. *Volta Review*, 85(3), 346–352.

Suggested further reading

Alpiner, J. G. & Vaughn, G. R. (1988). Hearing, ageing, technology. *International Journal of Technology and Ageing*, 1, 2, 126–135.

Brooks, C. P. & Newell, A. P. (1985). Computer transcription of handwritten shorthand as an aid for the deaf: a feasibility study. *Journal of Man-Machine Studies*, 23, 1, 45–60.

Espin, C. & Sindelar, T. (1988). Auditory feedback and writing: hearing disabled and non-disabled students. *Exceptional Children*, 55, 1, 45–51.

Exley, S. and Arnold, P. (1987). Partially hearing and hearing children's speaking, writing and comprehension of sentences. *Journal of Communication Disorders*, 20, 5, 403–411.

Kretschmer, R. R. (Ed.) (1985). Special Issue: Learning to write and writing to learn. *Volta Review*, 87, 5, 1–185.

Maxwell, M. (1985). Some functions and uses of literacy in the deaf community. *Language in Society*, 14, 2, 205–221.

Schilp, C. (1989). Correcting grammatical errors with MacWrite. *Volta Review*, 91, 3, 151–155.

Chapter 9

Writing, Computers and Visual Impairment*

Richard Ely

In the past ten years, microcomputer use in schools has grown tremendously; at the same time, strides have been made in adapting microcomputers for visually impaired individuals. Every few months brings a new hardware or software solution. Much attention has been given to the technology itself; however, the actually implementation of these devices to determine how they can best be used requires more consideration. A better understanding of the cognitive processes used in a particular domain, such as composing, will assist in the development of access tools that better meet the needs of users. Such an understanding may also help those who teach the use of computers to visually impaired people to help their students use these adaptive tools more efficiently. This article will briefly review some of what is understood about the composing process. It will then examine the implications of that process as it relates to the implementation of computers in writing programs for visually impaired users.

To accomplish these aims, it has been necessary to bridge a number of domains: cognitive psychology, composition, computer-based writing, and adaptive technology. References to hardware and software are included, but there is still a need to describe more practical implementations of these tools in school and work settings.

Composition and cognition

The cognitive process for composing is one that for the sighted writer places enormous burdens on short-term memory. Flower and Hayes (1980), in examining cognitive processes used by writers, have outlined four elements of the process: planning, generating, transcribing, and editing. Their work indicates that these elements of composing are not discrete stages. The writer does not plan, generate, and then transcribe. Instead, in their model, the writer may move from one to another of these processes as the task demands, which calls on short-term memory at every stage. It also indicates the importance of the writer's ability to immediately reference the already written parts of the text.

Short-term memory

Miller (1956) notes that, in writing, the average unit held in short-term memory is a six-word clause, and that once a clause is read or set down, the exact words begin to fade, though the meaning is retained in long-term memory. While this limited memory space is working to retrieve information from long-term memory, it is also being used to monitor text transcription, including grammar and mechanics. Short-term memory cannot be expanded beyond a certain capacity, but experienced writers can devise strategies to increase its efficiency.

The sight-impaired writer often has even greater burdens placed on short-term memory. Any means used to access text, such as large print or magnification, limits the area of text available for examination. Thus, the visually impaired writer will have to learn to keep more accurate track of the text already generated. In addition, manipulation of a magnifier or a closed-circuit television (CCTV) further complicates the reading-writing task. When braille is used, the writer must also use short-term memory to recall correct contractions.

In the past 15 years, interesting alternative methods for teaching composing have been developed. Many of these methods can help novice writers learn to focus on specific aspects of their writing, rather than trying to carry out all functions simultaneously. The result can substantially reduce the load placed on short-term memory. Such composing methods may help visually handicapped writers if their means of setting down text is adequate and if their generated text can be easily accessed.

Peter Elbow (1975, 1981) and Donald Graves (1983) both offer well-developed programs to assist the novice writer. Each approach involves having students work through a number of drafts of a text with virtually no attention being paid to grammar or mechanics in the early stages. Elbow encourages such approaches as free-writing as a means for writers to get their ideas down. The writer starts writing and does not stop for a set period of time. The text produced is then a starting point for further writing. Graves emphasizes publishing students' written work to establish the concept of writing for an audience and the importance of the teacher writing with students to provide an example of how a writer works.

These methods, or similar methods, are effective for visually handicapped writers but only if the ways used for creating text are efficient and if the text is accessible to them and others at all stages of the writing process. The writer who uses magnification may find the generation of text an arduous task. Generating text can so slow down a process like free-writing as to make it useless. Such a student may find the typewriter more satisfactory for generating ideas quickly, but this benefit will be limited by the difficulty of reading the typewritten text.

Braille

Writers who must rely on braille as the composing medium will also find many of the new approaches to writing difficult. The use of the conventional mechanical braille writer requires much more effort than does the use of the manual typewriter. Though no studies of typing versus brailling rates have been published, a group of experienced braille teachers agreed that a mechanical brailler would likely to be somewhat slower, given equal experience with both the braille writer and the typewriter. Text production would probably be faster if the writer were able to use either an electronic brailler or a refreshable braille device, such as a VersaBraille, since these tools require less physical action to produce cells. The writer generating text in braille will not be able to share the finished work with sighted peers without first producing a second copy in print. Text accessibility is important for the mainstreamed student, especially since many writing curricula now involve peer editing of drafts of compositions. If visually handicapped writers are to benefit from this method of composing, they must have a writing tool that allows easy production of text, convenient access, and a means to make the text accessible to others. An Apple II computer equipped with both a printer and a braille embosser could be used with BEX software to produce braille for the writer and to print copies for his or her sighted peers.

The editing process

Recopying a new draft has always adversely affected the willingness of students to edit their texts. In her work with sighted children, Colette Daiute (1985) found that 'most children report that they prefer writing on the computer because it is more fun and because they don't have to recopy'. The problem of recopying is of particular relevance to writers who produce rough drafts in braille and who know ahead of time that they will have to produce at least one final print copy from braille.

If the writer has a computer that provides alternative output, such as a voice synthesis, enlarged characters, or refreshable braille, the writer can use such approaches as free-writing, knowing that the generated text will be accessible later. If the alternative display is effective, it can also lessen transcribing problems, thus reducing the load on short-term memory. The writer will not have to give as much attention to monitoring a text for mechanical or grammatical errors if he or she knows that at any point the text can be easily reviewed and corrected.

A growing number of braille translation programs are available for use with microcomputers. Writers who have access to a braille embosser can use these software tools to produce both tactual copies for their own use and to print versions for their sighted readers. The ease of text access a given adaptive device can provide will directly affect the degree to which the cognitive load on the sight-impaired writer can be reduced.

The importance of text access

Research on available alternative methods for a written communication has been reviewed by Heinze (1986), who found that little attention has been paid to composing strategies used by blind and sight-impaired writers.

One of the few cases of an experienced blind writer is a study by Gere (1982). Her subject, Jackie, was in her early thirties, with a BA in English and an M.A. in educational psychology. Jackie's first experiences with reading and writing were in braille, since her vision loss was a result of retinopathy of prematurity. Over the course of the study, Jackie was asked to perform a variety of writing exercises using any tools she wished and taking as long as she wanted to complete the exercises. Gere found that her pattern for writing was substantially different from that of most sighted writers. Jackie spent an unusually long time planning what she wanted to write, and when she wrote, she produced a single draft at the typewriter. Most sighted writers would have spent less time planning and much more time generating and editing their text. Many experienced sighted writers would have developed particular approaches, such as outlining or free-writing, which would have assisted them in the writing process.

Gere studied Jackie with the intention of discovering if blind writers use techniques for revising that might be helpful to sighted writers. Since Jackie's text was generated in print, a medium inaccessible to her, all her steps in composing had to be compressed into the stage before text generation. Once her text was set down, it was nearly impossible for her to make changes or corrections. Gere found that Jackie was able to recollect a written paper in great detail, but she did not interact with her text the way sighted writers do. In only one of the four writing tasks did Jackie elect to work in braille, her only means of interacting with her text. She then translated that text to print.

Jackie's single-draft approach is quite typical of the approach used by experienced blind adult writers (Ely, 1986). It largely precludes the interaction with the text that Flower and Hayes found to be an important element in the probing of long-term memory, since it demands the development of a set of elaborate compensatory skills. These skills allow such blind writers to use their long-term memory in recalling information and in holding a detailed recollection of their text. Visually handicapped writers must also cultivate cognitive approaches for utilizing short-term memory, which permits them to shuffle information quickly in and out of long-term memory. These skills, however, are not easy to acquire.

A computer with an adaptive interface can provide writers with text access to reduce demands on short-term memory. Such a computer would also reduce the need for the elaborate cognitive juggling skills that talented blind writers seem to require.

Gere found applications for sighted writers in Jackie's approach to writing. For example, Jackie was able to refine her text draft in the planning stage. Gere, however, did not take into account the fact that Jackie did not have access to her text for later use unless she sought the help of a sighted assistant, nor did she address the problems that Jackie's method of writing introduces when text must be generated over an extended period of time, as in writing a book.

There is a need for research into the methods used by both experienced and novice blind writers. Such research could provide guidance in developing a writing curriculum for novice writers and might also indicate ways computers might be effectively integrated into such a curriculum. These issues are particularly important in teaching composing to visually impaired students in a mainstream setting. If regular computer access is not possible, then the teacher of the visually impaired student and the English teacher will need to assist the student in learning the cognitive skills needed to circumvent the constraints of generating in the print medium. The alternative is to compose in braille and then translate for the sighted audience. This approach may reduce the effectiveness of a writing program in which students are expected to work through a number of drafts of a composition. If student writers have access to a computer with an adaptive interface, much of the curriculum designed for use with sighted students will be applicable to them.

Adequate and effective adaptive computer tools for writing

Selection of computers with adaptive devices must be individualized. One must take into account the medium in which the individual is currently working; the speed and efficiency with which the individual uses that medium; any fluctuations in visual condition; and the task, in this case writing, the individual will be expected to carry out. There has been a tendency to prescribe computer systems as units, with minimal attention paid to evaluating individual needs and the variety of devices available that might meet these needs. The same attention should be paid to prescribing adaptive computer devices as is currently paid to prescribing other low vision devices. The process of selecting these devices is as complex as that of assessing the need for vision devices. In the case of writing, one must take into account the computer, the word processing software, other writing tools such as spell checkers, and the adaptive technology that replaces the conventional cathode ray tube (CRT) presentation. The developmental stage of the writer, and the weaknesses and strengths of the individual's writing, also must be assessed. It is important to find a system that will not require an inordinate amount of time learning how to use the computer before it can be used effectively as a writing tool. It is helpful if both the classroom teacher and the teacher of visually impaired students understand computers and their applications in writing. This is important both in the selection of the configuration of hardware and software and in its implementation for the writing program.

Device selection

An important criterion for selecting a composing device is the accessibility of the generated text. None of the alternative display formats currently available can simultaneously provide alternative display and the aggregate presentation of text afforded the sighted writer by scanning an entire page or screen of the text. Most refreshable braille display devices present less than one line of computer screen text at a time. Tools for enlarging computer display are restricted in the same way as optical magnification, that is, the higher the level of magnification on the screen, the less the amount of text viewed. Voice synthesizers present text in the same linear fashion as a person reading the text aloud, one word at a time, would.

When a teacher begins working with a student writer who is employing one of these devices, it is important to consider the differences in modes of display. Any access to text is preferable to none, but alternative display devices should be judged by the features that reduce access to text. In many cases, teachers and students must work with the system to devise approaches for text scanning to meet the individual needs of the student. Such software features as search, search and replace, and text marking can be valuable and may be part of the adaptive aid or of the application software.

All too frequently, adaptive device design does not allow for an effective display of language elements. This lack of attention to the use to which these tools will be put can be seen in the design of many screen review software packages developed for use with voice synthesizers. Most of these tools display a character, word, line, or screen of text. Word processors are the most commonly employed software tools used with these adaptive devices. Despite the fact that the most common language structures are sentences, few of these tools will allow the writer to read back written text in sentences. The result is that the screen most often displays defined lines of text, which breaks the natural flow of the writing. If these systems allowed the writer to review a sentence or a paragraph, they would be facilitating the process of composition, not inhibiting it.

Though adaptive devices are still expensive, one may wish to consider bimodal display, a combination of two alternative displays. Though no current studies explore the potential of such displays for writers, some writers have found multiple display advantageous. As an example, a given user may find it easy to track text as it is typed by using enlarged character display but find it difficult to gain a sense of the flow of the writing if the entire text must be read visually. A voice-synthesized presentation may make available more of the composition at a time, though it is often difficult or impossible to detect such problems as misspellings with speech. In this case, a writer may benefit from a combination of enlarged character display and voice-synthesized text.

Bimodal display may not be economically feasible for many people now, but it may be affordable later. The Apple Macintosh has a synthesizer chip built into the machine itself, and soon there will be software that permits the writer to obtain an interactive aural display of text. Berkeley Systems Design, the same company that is completing the voice software, has already marketed Inlarge, an enlarged character display tool. When complete, both pieces of software will run at the same time, offering two modes of display for the cost of an inexpensive synthesizer.

The Telesensory System Incorporated (TSI) Vista display system, which provides a magnified display of both text and graphics, now operates with a range of screen review tools used in conjunction with a voice synthesizer. Together, these devices permit two modes of display for the IBM and compatibles. More research is needed to establish the potentials for bimodal display for sight-impaired computer users.

Word processing

Selecting the correct word processing software for any writer is important, especially for the sight-impaired writer. Even more care must be taken if the writer is a student. A number of word processors, such as the Magic Slate and Talking Text Writer, have been designed for young writers. Both these tools allow for 20-, 40-, and 80-column display, which may assist low vision students. They offer simple functions for the beginning writer but are often inadequate for an older student.

Many of the newer, more sophisticated word processors can be used easily by young writers, and writing, editing, spell checking, and saving of text can be carried out with a few simple commands. One may want to consider selecting software that will serve the student in college and beyond. As a teacher, one does not want a student to have to spend a great deal of time learning to use new programs. The computer should be a tool for writers. It should not, because of poorly designed software, have to become a subject in itself.

It is important to know how well, if at all, a given word processor will function with the selected access technology. Many word processors utilize the features of the CRT screen. They may use colored text, highlighted text, and a range of dynamic displays that work for the screen reader but that may not convert to a suitable alternative display. One must make sure that all the functions of the word processor can be easily accessed. A writing aid that is hard to use for display may sit on the shelf.

For some writers, one of the word processors that has been designed with a particular alternative display in mind may be a good selection. These tools may be immediately usable, but they are nearly always limited by the time a software company has to take to develop them. The market for such tools is small and the cost of development will be borne by a much smaller group of users. This often means that there are many fewer editing and formatting features available.

It also may not be possible for users of this specialized software to share documents that have been written with other word processors. Word processors such as the Talking Screen are designed specifically for speech output. The design does not necessarily mean, however, that they will provide any better text access than would conventional word processors used with an adaptive interface.

Ideally, one should take the time to use any product at length before purchasing it. Regrettably, there are not enough access centres in the country that are organized to allow teacher and students to examine alternative display tools. Though there is a need for unbiased evaluation of adaptive devices and of those tools when used with particular application software, one must frequently rely on word-of-mouth reports or on the promotional material provided by vendors.

The American Foundation for the Blind's National Technology Center has carried out independent evaluations that have been published in the *Journal of Visual Impairment & Blindness*. User Network (AFB) also provides reliable information supplied by technology users. Anyone should feel free to call the American Foundation for the Blind for User Network information.

Software tools

Though more and more composing software tools are being designed for student writers, it is important to determine which, if any, of these programs might be useful to a visually impaired writer. As with the word processor, the needs of the writer, based on his or her strengths, and weaknesses, must be evaluated. It is also important to determine how well these tools will work with the alternative display device.

Spell checkers are a very common writing tool. Many of the better word processors include a spell checker as part of the software package. A simple checker compares all words in the text with a dictionary. It informs the user if a text word is not in the dictionary. The spell checker corrector, like the one found in WordPerfect, goes a step further and utilizes a lexicon program to examine text words that do not match dictionary entries and then selects a number of words from which the writer may choose. This type of tool can be helpful to a sight-impaired writer who has had far less exposure to the written language than have his or her sighted peers. As a result, the sight-impaired writer has not developed the word recognition skills of spelling that come with visual exposure to text.

Students of braille learn contractions that may make it difficult for them to recall complete spellings. Often, sight-impaired writers are aware of this shortcoming in their writing and work hard to monitor their spelling. This monitoring can overtax the cognitive resources they need to organize and generate ideas. A good spell checker can be used as another tool for reducing the cognitive load.

Prewriting tools have been developed to assist the writer in getting started with a writing project. Programs such as QUILL, marketed by D.C. Heath and Company, provide a variety of methods to assist the writer in the early steps of writing. The author is currently evaluating several of these programs to determine their usefulness with adaptive technology.

If these programs can be efficiently accessed with an alternative display device, they might also aid the sight-impaired writer in much the same way as they do the sighted writer. A growing number of text analysis programs, such as Grammar and Style, are being evaluated for adaptive use. These programs prompt writers to examine their own text more critically. For example, the program might indicate overused words, or it might display the length of sentences to encourage variety in sentence length. These programs may be as useful to visually impaired writers as they are to their sighted peers. These programs do not correct errors, however; the software only identifies textual elements the writer may wish to examine.

Some of these tools have elaborate systems to display their analysis. Thus, their usefulness to a writer using an alternative display may be problematic. As noted, if the tool requires an inordinate amount of time to employ, it is unlikely that the writer will use it frequently. It is unfortunate that there are no published studies on the effectiveness of any of these tools when they are being used with specific alternative display devices.

Conclusions

As computers are more widely employed as tools for composing, they will come to offer more advanced composing functions. If writers with limited vision are unable to use these powerful tools effectively, then sighted writers will have a distinct advantage based on technology alone. The word processor is quickly replacing the typewriter as the primary correspondence tool of business.

Thus, a lack of computer access may not only restrict the potential for academic performance for sight-impaired persons, it may also have an adverse effect on sight-impaired individuals as they seek jobs. But it is important to realize that the computer is not a panacea for all persons with impaired vision.

Computers cannot and should not be thought of as replacements for such devices as the slate and the stylus, nor will they eliminate the need for the development of strong compensatory skills. In many ways, the computer has added to the expectation of teachers of visually impaired students. Teachers must not only foster computer use but must also continue to teach their students the traditional skills needed to succeed in school.

Only a few states are currently providing the support needed to grant visually impaired students meaningful access to computer tools. In some schools, a student will be fortunate to have an hour or two of computer instruction a week. When it is appropriate, the teacher must work to try to include the acquisition of computer hardware and software as part of the Individualized Education Plan (I.E.P.). Ideally, students should have access to equipment that can be used both in their classrooms and at home. It is frustrating for a student to know that there is a tool that could make him or her more independent when that tool is virtually unavailable.

As teachers, we must help parents to see the benefits of these computer tools and, if it is economically feasible, advise them to consider purchasing adaptive equipment. For teachers to provide helpful advice, it is necessary that they keep abreast of developments in the field of computer access, seek out regional resources, and strive to gain working experience with these new tools.

One may argue over computer use in elementary and secondary schools in the future. Some schools have excellent programs for computer implementation; other schools offer virtually no computer instruction. It is becoming clear that computer technology and adaptive tools are at a stage of development in which they can be of inestimable value to sight-impaired students. It is important for these students to develop the knowledge and skills they will need to make these devices work for them. The task environment of composition is but one example of the ways in which the practical application of computers can facilitate learning.

Teachers of visually impaired individuals must work with classroom teachers to examine other disciplines in which computers may facilitate their students' learning. There is an additional need for information to guide teachers as they work to employ these tools. This information may also assist developers of adaptive interfaces to produce more effective tools. If these issues are ignored, computers will create an insurmountable barrier for visually impaired students: with careful thought, research, and a willingness to experiment, however, computers may become liberating educational tools.

References

Daiute, C. (1985). *Writing and computers*. Reading, MA: Addison-Wesley.

Elbow, P. (1975). *Writing without teachers*. New York: Oxford University Press.

Elbow, P. (1981). *Writing with power*. New York: Oxford University Press.

Ely, R. (1986). *Typing it in, talking it out: A formative evaluation of the talking word processing software*. Unpublished doctoral dissertation, Harvard University, Cambridge, MA.

Flower, L. S. & Hayes, J. R. (1980). The dynamics of composing: Making plans and juggling constraints. In Gregg, L. W. & Steinberg, E. (Eds.), *Cognitive processes in writing*. Hillsdale, NJ: Lawrence Erlbaum Associates.

Gere, A. R. (1982). Insights from the blind: Composing without revising. In Sudol, R. A. (Ed.), *Revising: New essays for teachers of writing*. Urbana, IL: ERIC Clearinghouse on Reading and Communication Skills and The National Council of Teachers of English.

Graves, D. H. (1983). *Writing: Teachers and children at work*. Exeter, NH: Heinemann.

Hayes, J. R. & Flower, L. S. (1980). Identifying the organization of writing processes. In Gregg, L. W. & Steinberg, E. R. (Eds.), *Cognitive processes in writing*. Hillsdale, NJ: Lawrence Erlbaum Associates.

Heinze, T. (1986). Communication skills. In Scholl, G. T. (Ed.), *Foundations of education for blind and visually handicapped children and youth*. New York: American Foundation for the Blind.

Miller, G. (1956). The magical number seven, plus or minus two: Some limits on our capacity for processing information. *Psychological Review*, 63, 31–97.

(Suggested further reading is given at the end of the next chapter)

Chapter 10

Developing Writing and Word Processing Skills with Visually Impaired Children: A Beginning*

Alan J. Koenig, Catherine G. Mack, William A. Schenk and S. C. Ashcroft

The microcomputer explosion is being felt with increasing magnitude throughout the entire educational system, and our field of visual impairment is feeling much more than just slight tremors. Scadden (1984) states insightfully that 'The microprocessor . . . is shaping a changing civilization' (p.394) and proposes that its application has the exciting potential to equalize skills of visually handicapped and sighted individuals in the new information age. If the potential for achieving equality exists, our challenge is to make it a reality. This report describes a beginning effort to meet such a challenge.

George Peabody College for Teachers of Vanderbilt University and Tennessee School for the Blind (TSB) have conducted several joint endeavors involving microcomputers and access technology. High school students at TSB have participated in a computer club sponsored by Peabody college, and students have served as subjects in 'Research on multimedia access to microcomputers for visually impaired youth,' a federally-funded Peabody project. Our present project emphasizes microcomputer word processing, a natural extension of this prior research.

In the spring of 1984, the Apple Education Foundation (AEF) awarded a two-year joint project to Peabody College and TSB. Its purpose is to examine a major application of microcomputers: the development of writing and word processing skills with visually impaired students. The project provided us with Apple IIe microcomputers, ink print printers, and software. Donations of special technological aids came from other sources.

Three aspects of our project will be discussed in this article: The conceptual framework, activities at the project site, and an instructional module for introducing students to the Braille-Edit word processing program.

Conceptual framework

Children must develop four communication skills in order to function effectively in a literate society: speaking, listening, reading, and writing. Today, many believe that computer skills must be added to this list. Given appropriate opportunities, visually impaired children usually develop good speaking and listening skills, but blindness may erect a more severe barrier to the learning of reading and writing skills.

Although braille affords a blind reader a great deal of independence, it also imposes certain limitations on reading and writing. Braille reading is inherently slower than print reading because of the sequential nature of tactile perception and the perceptual unit's restriction to a single braille character. Given the relatively slow rate of tactile perception, as well as the bulk and cumbersomeness of braille and the restricted availability of braille reading materials, blind children generally read less than their sighted peers. Thus, writing skills may be impoverished because these students spend less time interacting with printed (brailled) materials.

A more fundamental limitation basic to reading and writing can occur when blind children are limited in the range and variety of life experiences. Without appropriate background experiences, development of concepts and schemata is hindered and growth in reading and writing is restricted.

However, given early intervention to provide a rich and varied range of opportunities for experiences and qualified teachers to foster positive attitudes and provide a sequential instructional program, blind children can develop good braille reading and writing skills in accordance with their abilities.

Braille writing, nevertheless, continues to involve basic difficulties. It is slow, and errors are difficult to correct. Revision of an initial draft of a manuscript is restricted due to the mechanics of writing braille on paper. There is no convenient method for teachers to provide timely corrective feedback. Thus, ideas which children wish to communicate in writing may be trapped within them because the Braille writing mode is a barrier to free expression.

Braille word processing holds the potential to reduce this barrier because it presents the writer with a more efficient way to communicate ideas. Word processing systems designed for braille readers allow blind students the freedom to write and then easily to make revisions and corrections before a final copy is prepared. Never before have blind persons had this option which could more nearly approximate the advantages available to those of their sighted peers.

Report on project activities

Equipment: The AEF project provided us with 30 complete sets of Apple II microcomputer equipment, including monitors, CPUs, and duo-disk drives. Most of these were placed in classrooms, dormitories, and the library of TSB to enable our students to have access to microcomputers during the school day, after school hours, and on weekends. Two sets of equipment placed at Peabody College are used to develop instructional materials and give preservice training. We also received ink print printers, modems, Apple Write software, and

Apple Logo software. Through AEF we obtained Bank Street Writer software and Echo II speech synthesizers.

We also received gifts of technological equipment from several distributors for use in our project. Maryland Computers donated a Cranmer Modified Perkins Brailler and a Total Talk Computer Terminalk. Visualtek gave a Large Print Computer to the project. To complete the list of equipment resources, TSB purchased multiple copies of the Brailler-Edit word processing program and additional sets of the Echo II speech synthesizer. All equipment is currently in use by students and their teachers.

Learning goals and activities

The project has two major goals. The first relates to teachers. They are expected to become proficient with basic procedures of using microcomputers and access technology for visually impaired pupils as well as mastering the use of Braille-Edit and either Apple Writer or Bank Street Writer. A 35-hour workshop for teachers and administrators at TSB was conducted during the summer of 1984 by project assistants from Peabody College and TSB for the purpose of achieving this goal. Teachers who entered this workshop with little or no knowledge of microcomputers left with remarkable achievements to their credit. A follow-up 'refresher' workshop was held for one day before the beginning of the school year, and a short introduction was given to those who could not attend the summer workshop.

Only seven months into our project, teachers are not only putting into practice information gained from the workshop, but also are looking for and developing other applications for use of microcomputers and word processing. One teacher has pursued and obtained through a local computer shop a means of generating large print output from a dot matrix printer. Another teacher has made study guides for the Tennessee Proficiency Test available in braille using Braille-Edit and the Cranmer Modified Perkins Brailler.

Another component of our first major goal was to introduce preservice teachers at Peabody College to microcomputers and word processing for blind students. These vital skills were incorporated into the introductory course on Braille Reading and Writing during the fall of 1984. Several preservice teachers volunteered additional time to instruct students in word processing at TSB during the fall semester. As a result, six TSB students reached a level of word processing proficiency which included creating, editing, saving, and printing their work with only a minimal amount of supervision.

The second major goal of our project relates to the development of writing and word processing skills by visually impaired students at TSB. Before beginning instruction in word processing skills, all students are expected to master basic computer skills such as knowing the parts of a microcomputer, loading and booting disks, running a program, and shutting down the computer. Then braille readers are introduced to Braille-Edit and print readers are usually introduced to Apple Writer or Bank Street Writer. A subgoal of our project is to develop instructional materials for use in teaching word processing skills to visually impaired students. A module for introduction of Braille-Edit is described in the next section.

Students at all grade levels have begun learning to use the microcomputers. First through third graders learn names and purposes of each piece of equipment, and how to run simple programs. Students in the fourth, fifth, and sixth grades complete a short computer literacy unit and are introduced to word processing. Several fourth graders have started using the Braille-Edit program for such applications as writing their weekly spelling sentences with word processing. Fifth and sixth grade students have become quite proficient with Braille-Edit and are required to complete two written assignments per week with the microcomputer, although they often produce more.

High school students have also begun to apply their computer and word processing skills. Several have achieved such a high level of mastery in word processing that they have been able to join an after-school work experience program producing workbooks and worksheets in braille. One high schooler has earned extra money using Braille-Edit to transcribe menus into braille for local restaurants, and the student government has used Braille-Edit to braille versions of a pamphlet for the Blood Drive and a CPR manual.

In order to examine the development of word processing skills and their impact on writing skills, we collect data from students' writing samples, including samples written with a braille writer, slates and stylus, handwriting, and word processor. Results from the baseline samples and first retesting will be described in detail in an upcoming article.

Instructional module for braille edit

An integral part of our project is the development of instructional materials for introducing word processing skills to visually impaired students. The major components of a module for instruction in Braille-Edit are described below.

Braille-Edit is a powerful, complex word processing system developed and distributed by Raised Dot Computing Company. It contains two disks and is run from four menus. The typical user, and especially school children, will most likely not have a need to use many of its features. However, given appropriately sequenced instruction in selected features of Braille-Edit, blind students can achieve the skills necessary to complete a wide variety of writing tasks. Although Braille-Edit allows the user to enter information in a variety of ways (e.g., VersaBraille, Perky, braille keyboard on Apple computer), we chose to emphasize a common method which uses the regular keyboard for input with Echo II as an access device and hard-copy braille and print for output.

The commands that would be most useful to students were selected and incorporated into an instructional module. We selected six items from two menus that allow students to create and save a chapter, translate into Grade 2 braille, reverse translate from braille to print, print a chapter, and obtain a disk catalog. Also, we presented instruction on inserting and deleting text by the character or word. In order to review a file and then resume working, the cursor commands for advancing to the end of a chapter and for returning to the beginning were included. The most vital commands for use with Echo II enable students to hear each keystroke as they are typing, to cancel this option, and to read large portions of the text in a word-by-word manner. Students also review character by character, using the right and left arrow keys. Finally, four formatting commands were presented in the module, allowing students to set braille or print page numbers, double space, and center text.

Figure 10.1. Checklist of Essential Procedures and Commands in Braille-Edit 2.50

Student's Name .... Observation Dates .... Mastery

Basic Procedures	**Yes**	**No**
Loading and booting the disks	[ ]	[ ]
Entering configuration	[ ]	[ ]
Moving to the main disk (J)	[ ]	[ ]
Creating a chapter (E)	[ ]	[ ]
Saving a chapter (Control Q)	[ ]	[ ]
Grade two translation (G)	[ ]	[ ]
Reverse translation (B)	[ ]	[ ]
Printing a chapter (P)	[ ]	[ ]
Getting a disk catalog (D)	[ ]	[ ]
Insert and Delete Commands		
Insert text (Control I)	[ ]	[ ]
Delete single characters (Control D)	[ ]	[ ]
Delete x number of characters (Control D,#)	[ ]	[ ]
Delete single word (Control DW)	[ ]	[ ]
Delete x number of words (Control D,#)	[ ]	[ ]
Cursor Movement Commands		
Advance to end of chapter (Control A)	[ ]	[ ]
Move back to beginning of chapter (Control Z)	[ ]	[ ]
Echo II Commands		
Announce all keystrokes (Control SA)	[ ]	[ ]
Cancel announce all keystrokes (Control SA)	[ ]	[ ]
Read text (Control O)	[ ]	[ ]
Formatting Commands		
Braille page numbers ($$nb)	[ ]	[ ]
Print page numbers ($$np)	[ ]	[ ]
Center text ($$c)	[ ]	[ ]
Double spacing ($$12)	[ ]	[ ]

Comments: ....

....

....

....

The instructional module contains a thorough description of procedures, as well as the exact words spoken by Echo II throughout the sequence of steps. Also included is a 'Checklist of essential procedures and commands in Braille-Edit' which can be used as a checklist for evaluating students' mastery of the selected features. This checklist is presented in Figure 10.1 and may assist teachers who are just beginning to teach the Braille-Edit system to their students. Finally, the module includes a 'Braille-Edit help chart' which outlines basic steps in the essential procedures and commands. This chart is presented in Figure 10.2. It is intended for students to use as a quick reference guide after

Figure 10.2. Braille-Edit 2.50 Help Chart

Writing Something
1. Boot BOOT disk
2. Enter configuration
3. J-Jump to Main Menu
4. Insert MAIN disk
5. E-Editor
6. Enter text
7. Control Q-Quit and Save

Translating Print into Braille
1. G-Grade 2 Translator
2. Enter chapter name and return
3. Return again
4. Enter braille chapter name and return

Translating Braille into Print
1. B-Back from Grade 2
2. Enter chapter name and return
3. Return again
4. Enter print chapter name and return

Printing Something
1. Get printer set
2. P-Print
3. Enter chapter name and return
4. Return again
5. Enter printer number
(from configuration) and return

Commands Used in Editor
1. Control I-Insert
(control I, enter text, control N)
2. Control D-Delete
(control D, enter number, space bar)
3. Control DW-Delete word
(control D, enter number, control W, space bar)
4. Control A-Advance to end of text
(control A, space bar)
5. Control Z-Return to beginning of text
(control Z, space bar)
6. Control SA-Announce all keystrokes
(repeat to cancel)
7. Control 0-Read text with Echo II
8. $$c-Center text
9. $$12-Double spacing
10. $$nb-Number braille pages
11. $$np-Number print pages

they have completed the module and does not take the place of the module itself.

When students achieve mastery of essential procedures and commands, they are ready to learn additional features of Braille as their need arises. The Braille-Edit manual and reference card presented in braille from Raised Dot Computing which accompany the software will be useful in these situations. We have found, however, that features outlined in Figure 10.1 will allow students to use Braille-Edit to accomplish most writing tasks.

A brief look at the future

This article has described the beginnings of our effort to examine the development of writing skills and word processing skills with visually impaired students, made possible through a grant from AEF. In the upcoming months, we plan to continue gathering information on students' word processing skills with a focus on how these skills impact on their writing skills. We will also be revising and extending our instructional materials for teaching word processing skills, including the use of print programs such as Bank Street Writer. It is our hope and intention that the outcome of this project will help us meet the challenge to bring equality to visually impaired students in the vital area of written communication through the use of microcomputers and word processing.

Suggested further reading

Edwards, A. D. (1989). Soundtrack: an auditory interface for blind users. *Human Computer Interaction*, 4, 1, 45–66.

Hartley, J. (1989). Text design and the setting of braille: with a footnote on Moon. *Information Design Journal*, 5, 3, 183–190.

Hartley, J. (1990). Author, printer, reader, listener: four sources of confusion when listening to tabular/diagrammatic information. *British Journal of Visual Impairment*, 8, 2, 51–53.

Maley, T. (1987). Moon à la mode. *New Beacon*, LXXI, 840, 109–113.

Mercer, D., Correa, V. L. & Jowell, V. (1985). Teaching visually impaired students word processing competencies: the use of Viewscan Textline. *Education of the Visually Handicapped*, 17, 1, 17–29.

Parkin, A. J. & Aldrich, F. K. (1989). Improving learning from audiotapes—a technique that works. *British Journal of Visual Impairment*, 7, 2, 58–60.

RNIB (1990). *Computers at Work*. Royal National Institute for the Blind, 224 Great Portland Street, London W1N 6AA.

Chapter 11

Senior Citizens
Writing for Posterity*

Sydney J. Butler

Introduction

'Writing for Posterity' was a course in life-writing offered at a social centre for senior citizens in Vancouver, Canada, with the aim of helping the participants to write stories from their own lives. The project began with an illustrated talk that I gave to the members of the Brock House Society for seniors on the topic: 'From Cavepainting to Computers; How We Teach Children to Write', in which I tried to show how human history begins when early man records in the art of cave-painting the preoccupations of his life and culture. Using samples of children's writing collected from various schools, I showed how students could become involved and engrossed in their writing if they too felt that it served to record their interests and experiences, especially if they felt that other people were going to share their thoughts and feelings.

As several of the members of the Society expressed interest in this approach to writing, we decided to run an experimental course in which members would meet with the instructor once a week for ten weeks to 'write for posterity'.

Programmes for older adults

There are several precedents to suggest the value of such a programme. In a wide-reaching programme developed at Michigan State University, faculty and graduate students met seniors in residential and daytime care centres to promote the art of 'lifewriting' in group poems and autobiographical writing (Gillis & Wagner, 1980). Other projects in New York (Wright, 1981), Amhurst, Massachusetts (Bouchard, 1979), and Austin, Texas (Staples, 1981), all testify to the quality of expression that can spring from such classes.

But more important than the written products themselves is the effect that the writing can have on the participants. In the Michigan project the participants became more attentive to each other and more interested in current events and topical concerns. Dreher (1980) repeated the claim of Robert H. Butler, Chairman of the National Council on Aging, that the art of 'life-review' helps the writer to resolve conflicts and regrets and to reconcile personal relationships. One of the goals of the Michigan project was to use life-writing as a means by

---

* *Reprinted from* Adult Education, *1985, 58, 234-240, © National Institute of Adult Continuing Education. Reprinted with kind permission of the author and the National Institute of Adult Continuing Education.*

which elderly people could re-establish their feelings of self-worth and psychological well-being.

Writing for publication

One component of the writing course common to all of these projects is the necessity to provide opportunity for the participants to share and publish their writings. In my initial presentation I had shown how modern technology in the form of the micro-computer, now being used in some schools, enables students to draft, edit, and publish their writings in formats which approach the quality of traditional typesetting without the costs involved in commercial publication. While the projects mentioned above show the value of publishing the products from their writing courses in anthologies, magazines, and newsletters, most of which demand outside funding, I wanted to make the Writing for Posterity course a strictly do-it-yourself project which could be self-sustaining, rather than being dependent on any outside agency for financial support.

Consequently, during the first meeting I proposed that one of the aims of the course would be to have everyone produce a piece of writing which could be shared with the class and which could be printed by computer for duplication in a class booklet, and I showed them examples of such booklets produced by school students. A more long-range goal was to show the participants how they could themselves preserve their writings in such a form, and that if they continued to write and edit their ideas they could each produce an individual collection of personal stories which would be valued by their friends and family. Hence, the title: 'Writing for posterity'.

Beginnings

Until the first meeting I had no idea what skills, attitudes, and aspirations the participants would bring. I had outlined a programme expecting that they would all be literate, but probably not accustomed to any sustained writing. I assumed that these older writers would need help in generating ideas, encouragement in sharing their ideas, and some technical advice in the shaping and refining of their writings for publication. I felt that my first aim should be to convince them that everyone was a potential author.

At the first meeting there were fifteen people who had responded to the notice in the Centre's Newsletter. Of these, twelve maintained a consistent attendance with the occasional absence due to illness. The other three dropped out after one or two meetings. One other person joined the class for the final session, showed a book of her own writings and art work, and was immediately recruited to design the cover of the class book.

Generally, the level of ability far exceeded my expectations. All of the regular participants were women in their sixties and seventies. Many had retired from such professions as nursing, social work and teaching. What was most surprising was that many of them were very accomplished and successful writers. Several had been accustomed to writing as part of their professional activities; two others had had experience in writing and presenting scripts for radio broadcasts, another was a published author of children's stories, and one had actually published her own autobiography. It was clear from the first meeting that they had not come to the class to be taught to write. Instead, I proposed that the class would function as a support group, and that my role was more in the nature of a co-operating editor rather than as an instructor.

Nevertheless, I did persevere with some simple heuristics, and although one member mentioned that at first she thought these 'tricks' were 'Mickey Mouse' ideas, later she confessed that they did help her to generate ideas and collect details of the experience that she was describing.

We began slowly, exploring various techniques for reconstructing the experiences that have formed our lives. We made word-banks of places that are important to us, and then lists of words and images to make those places come alive. Each of the participants talked with a partner about the particular place she had chosen, while the partner asked questions to clarify the meaning of the experience. Each class culminated with a quiet writing-time when everyone, including me, wrote in an attempt to get the ideas down with a minimum of self-censorship, aiming to capture some of the fluidity of talk in the forced flow of writing.

Developing writing confidence

Our first exploration became known as 'The Sense of Place', as the participants found it convenient to give such labels as 'Dangerous Moments', 'Snapshots', 'Turning Points' to each of our writing experiments. For example, 'An Incident in Childhood' followed the reading of Countee Cullen's poem 'Incident' which proved to be a powerful stimulation of long-held memories, and resulted in a tremendous sharing of childhood experiences.

These introductory activities were designed to create a feeling of security among the group. Some had experience of other writing classes in which they had been given precise instructions as to how and what they were to write, and had the expectation that their writing was to be handed in to be marked by the instructor. But with this class the first four weeks went by without any critical examination of their writings. On the other hand, there had been a great deal of talk about their experiences, a lot of sharing of common memories, and a sense of trust developed among the members of the group. In fact, it became so easy to get them talking about their experiences that it was sometimes difficult to get them to shift into a writing mode. But as the weeks passed they were beginning to accumulate a hoard of personal memories and stories in their writing-folders and note-books.

Evaluating stories

For the fifth session I had asked each member to bring the draft of one of the stories as a 'work-in-progress'. In the previous week each member of the group had surveyed her stock of possible stories and the support of other members of the group helped each person to select a story for sharing. For the first sharing I structured the class in two groups, and had the members write their comments on separate response-sheets, as the stories circulated around the group. Beforehand, we had talked about the doubts and uncertainties of being an author, and discussed the type of feed-back that was best given at this stage. To ensure that the written responses would be positive, I insisted that everyone should begin their comments by completing two sentence-openers: (1) 'This is a good . . .' and (2) 'I like the . . .'.

Each writer thus received five or six independently-written comments, all of them in a positive and encouraging tone, pointing to the value of the experience that was narrated. The next stage was to have volunteers read a story to the whole group, while the listeners made written comments and questions for the author.

At this point in the course, I had to miss the next three sessions. But the class had already gathered momentum; they had confidence in presenting their writing to a sympathetic audience, and so they decided to continue the meetings during my absence. After a certain amount of unfocused discussion, they elected a chairperson to manage the discussion and arrange the sequence of readers, and the class continued successfully.

Publication

Throughout all the sessions I had dangled the carrot of publication before these authors, and I asked those who were satisfied with their stories, after receiving praise and questions from the rest of the group, to give me the stories ready for word-processing. Some of the authors submitted several stories, others worked with a partner to help them choose which story to submit. Again, there were faint shadows of apprehension in handing in a story, as if they felt they were back at school and handing in assignments to be marked.

However, I had promised that their stories would be transcribed as faithfully as a fast typist could put them into the computer, and that neither I nor the typist would make any editing changes. Instead, each of the participants received a clean print-out of her story. For the next few sessions the class encompassed a variety of activities. We still tried to devote some time to writing more stories, some of the time was devoted to oral presentation and discussion of how stories might be improved structurally. Other members worked with a partner, going over their printed stories to shake out any awkward phrasing. The revised stories were returned to the typist.

During the last two sessions the group's efforts were totally directed towards the publication of the class book. The members discussed and voted on such editorial concerns as how many stories were to be selected (one from each member), how the stories were to be arranged (in alphabetical order), what size and design of type-face should be used, how the booklet should be copied and bound, and how many copies were to be produced. Originally I had suggested that we would be producing only one copy for each member, but so many decided that they wanted extra copies for friends and relatives that in the end we had a hundred copies printed, the extra ones being sold to the other members of the Brock House Society at a price which covered the costs of typing and duplication. Finally, during the last session the class chose two teams of proofreaders to check the print-outs and also to choose standard formats for the titles and by-lines.

Perceptions about life-writing

During one of these working sessions the class, in my absence, decided to write evaluative comments about the course, and what they had learned from it. All were appreciative of the help that they had had in getting started. Several confessed to their fear of receiving harsh criticism and mentioned the sense of trust that had been built up slowly over the weeks with appreciative comments and gentle suggestions for improvement. Some felt that their writing had improved because of their attention to finding concrete details; sharpening beginnings and eliminating unnecessary explanations. Yet the main aim of the course was to enable the participants to recapture experiences, as one person commented: 'The course has left me wrung quite dry—purged of inner thoughts and dreams. I have felt in turn sad, happy, amused and sometimes quite startled at what emerged in the wee small hours of the morning'.

Conclusions

The culmination of the course was the publication of the class booklet, forty-four pages of text photo-copied on 11 x 17 inch sheets, back and front, and folded and saddlestitched in a cardboard cover. The computer word-processor allowed us to use combinations of normal, bold-face, and italic type, which, when justified, gave the resulting booklet the look of professional printing, but at a very moderate cost.

The most important result of the appearance of the booklet was that it made all of the participants realise that they had become published authors, and in their evaluative comments they showed that they wished to continue this process of life-writing.

When the class met again for the spring of 1985, the members expressed their determination to continue, but with the achievement of their first book behind them, they raised their sights to aim further afield. I had already suggested that with the group's abilities in organisation and administration, they no longer needed an 'instructor'. Instead they decided to organise themselves as an autonomous group within the Brock House Society, following the model of other groups such as the Bridge Club and the Choir. A meeting with the Executive Co-ordinator of the Centre raised the possibility of their applying for funding from the Federal Government's 'New Heritage' programme, diverting some of the group's writing energies towards the writing of a funding proposal.

This proposal led to the group's increasing self-awareness of their achievement, and in particular an awareness of how the group's success could serve as a model for the establishment of similar programmes in other senior centres and residential homes. Consequently, the proposal was formulated, not merely to support the continuation of the course as it had first been run, but to expand the publication to make it an outlet for other senior writers to publish for their peers. An expanded booklet would obviously be more expensive, but could accommodate suitable pieces of writing submitted by individuals from other centres who would learn about this group's efforts from newsletters and bulletins.

Another encouraging suggestion was that some of the members of the group would make use of the interviewing techniques practised in the class to collect oral history from other seniors in the centre or in the community, for editing for the publication.

Even more daring was a suggestion that some of the members would be able to use the class techniques to encourage life-writing in some of the local residential homes. The discussion that revolved around these and other ideas was an adequate confirmation that the participants in the class had evolved out of their initial 'student' role, and now they were ready to become their own 'teachers'. Becoming involved in the management of the class, taking the class attendance, collecting the money to fund the publication of their book and making arrangements to sell the extra copies at the Centre had built their confidence in their ability to manage their own affairs.

Whether the funding application to the government is accepted or not is immaterial. In making their proposal the members of the class had come to realise how they are able to make use of their own talents for organising and managing their own class, using their own resources to give help and encouragement to their fellow writers.

Achievements of the course

The course can be seen, therefore, to have achieved three distinct levels of success. Simply by providing a forum for discussion about writing and the sharing of the results, the class encouraged the participants to search their memories for their experiences. As one member commented: 'The classes have started people writing who have never written before—or have not written for some years'.

The class also helped the members to improve their writing, by learning to select and prune their ideas. Most of the criteria for success at the editing stage developed from the preparations for publication, and the class book, *Recollections in Writing*, provided both the motivation for improving writing and the evidence of the success of the project.

But the lasting achievement was in the class's determination to continue to function as an autonomous group, with the ability to become self-sustaining, to expand its activities and to provide leadership for other similar groups.

References

Bouchard, Lois Kalb (1979). Grey Heads, Not Grey Voices. *Teachers and Writers*, 10:3, 14–17.

Dreher, Barbara B. (1980). Directing a Writing Program for Retirees. *English Journal*, 69:7, 54–56.

Gillis, Candida, and Wagner, Linda (1980). Life-Writing: Writing Workshops and Outreach Procedures. Paper presented at the Annual General Meeting of the American Educational Research Association. (ERIC Document ED 186824).

Staples, Katherine E. (1981). A Writing Course for Elders: Outreach, Growth, Synthesis. Paper presented at the Community College Humanities Association Conference. (ERIC Document ED 212335).

Wright, Jeffrey Cyphers (1981). Tapping the Legacy. *Teachers and Writers*, 12:3, 31–37.

Suggested further reading I: Writing for the elderly

Czaja, S. J. (1988). Microcomputers and the elderly. In Helander, M. (Ed.) *Handbook of Human-Computer Interaction*. Amsterdam: Elsevier.

Meyer, B. J. F., Young, C. J. & Bartlett, B. (1989). *Memory Improved: Reading and Memory Enhancement Across the Life-Span through Strategic Text Structures*. Hillsdale, NJ: Erlbaum.

Rice, G. E., Meyer, B. J. F. & Miller, D. C. (1989). Using text structure to improve older adults' recall of important medical information. *Educational Gerontology*, 15, 527–542.

Taub, H. A., Sturr, J. F. & Monty, R. A. (1985). The effect of underlining cues on memory of older adults. *Experimental Aging Research*, 11, 225–226.

Vandenplas, J. M. & Vandenplas, J. H. (1980). Some factors affecting legibility of printed materials for older adults. *Perceptual and Motor Skills*, 50, 923–932.

Walmsley, S. A., Scott, K. M. & Lahrer, R. (1981). Effects of document simplification on the reading comprehension of the elderly. *Journal of Reading Behavior*, 13, 237–248.

Suggested further reading II: Writing by the elderly.

Butler, S. J. (1988). *Life Writing: Self Exploration and Life-Review Through Writing*. Dubque, Iowa: Kendall Hunt.

Czaja, S. J. *et. al.* (1989). Age related differences in learning to use a text editing system. *Behavior and Information Technology*, 8, 4, 309–319.

Supiano, K. P., Ozminkowski, R. S., Campbell, R. & Lapidos, C. (1989). Effectiveness of writing groups in nursing homes. *Journal of Applied Gerontology*, 8, 3, 382–400.

Part IV

Writing for electronic text

Chapter 12

Textual Cues as Variables in Computer-Based Instructional Screen Design*

Gary R. Morrison, Steven M. Ross, Jacqueline K. O'Dell and Charles W. Schultz

Since the early 1970s, there has been a growing interest in the field of instructional design with the layout and design of instructional text (e.g. Hartley, 1985). More recently, an interest has grown in applying similar typographical principles to the design of the computer-based instruction (CBI) screens (e.g. Grabinger, 1983; Hartley 1987). Not all computer displays, however, have a direct corollary with print-based typographical variables such as italics and boldface styles. Typical computer displays are (a) limited to 40 or 80 columns in 24 rows, (b) use only one font, (c) limit interline spacing to single, double, or triple spacing, (d) provide only a monospaced font. Many of these limitations, however, are overcome when the computer display uses bit-mapped graphics (e.g., Apple's Macintosh computers) which provide the designer with a wider range of fonts, sizes, and styles. These limitations, however, restrict the number of cues the designer can access when designing CBI displays for most computers. This chapter will summarize five studies we have completed on CBI screen design. The first two studies summarize our findings on text density while the remaining three studies discuss learner preferences for varying screen densities.

The computer screen is often compared to the printed page and critiqued in terms of the attributes of the printed page (e.g., Bork, 1987; Feibel, 1984; Grabinger, 1983; Lancaster & Warner, 1985; Richardson, 1980). In addition, current guidelines for screen design are often based on principles derived from research on printed instruction (e.g. Bork, 1987; Hartley, 1987) even though the two are recognized as different forms of presentation and have different strengths and limitations. It appears important then, that researchers and CBI designers pursue different alternatives to screen design than simply those principles generated from research on text-based materials to fully utilize the potential of the microcomputer.

Research on CBI screens design

Prior research on CBI screen design has either emphasized the manipulation of typographical variables or the manipulation of the content. The general con-

---

* *This paper has been specially prepared for this volume.*

sensus of researchers focusing on the manipulation of typographical variables is that screen displays should use liberal white space, double spacing, a standard ASCII typeface, and left-justified text (Allessi & Trollip, 1985; Bork, 1987; Grabinger, 1983; Heines, 1984; Hooper & Hannafin, 1986). Most 8-bit and 16-bit microcomputers severely limit the application of the typographical conventions (e.g., italics, boldface, underscore, leading, and justification) used in print, making applications and research using those principles very difficult if not impossible to implement. Research on manipulation of content by chunking the material into meaningful thought units which are then displayed on separate lines has failed to show clear advantages under either print or CBI (cf. Basset, 1985; Carver, 1970; Fiebel, 1984; Frase and Schwartz, 1979; Gerrel & Mason, 1983; Hartley, 1980; Keenan, 1984; O'Shea & Sinclair, 1983). Chunking does not change the amount or type of content presented, rather it changes the format of the presentation. It seems apparent then, that designers and researchers need to investigate other variables which can influence CBI screen design.

In the following pages, we will summarize our research on two additional screen design variables. First, we will describe a construct we have labeled 'text density' which concerns the *informational context* of the material presented. Second, we will discuss variations of the amounts of information presented on the screen, a variable we call 'screen density.'

Text density as a contextual variable

The first two studies on text density tested the effectiveness of an alternative method for displaying text on the computer screen. Unlike chunking which simply changes the display format of the information, the text density strategy manipulated the richness or detail of the information. Reder and Anderson (1980, 1982) used a similar methodology to compare complete chapters from college textbooks to summaries of the chapters' main points on both direct and indirect learning. They found the summaries to be comparable or superior to the full text in the ten reported studies. Their suggestion was that the summaries helped readers focus their attention on main ideas rather than on the additional elaborations provided in the full text. Although the research methodology used by Reder and Anderson has been criticized (Duchastel, 1983; Sherrard, 1988), the idea of reducing the amount of content presented is appealing, particularly in CBI. Specifically, we were interested in the level or richness or detail presented in the instructional text, a concept we have labeled as 'text density'.

As defined, text density was manipulated by varying the (a) length of the materials (e.g., word count), (b) redundance of ideas, and (c) depth of conceptual support for important concepts. Text density is similar to what reading researchers have labeled 'microstructure' (Davidson & Kantor, 1982). These two terms are contrasted with the concept of macrostructure which is concerned with how information is organized and elaborated through comparison of examples, nonexamples, and concept categories (DiVesta & Finke, 1985; Frayer, Frederick, Klausmeir, 1989, Moes, et al., 1984; Reder, Charney, & Morgan, 1986). In practice, the macrostructure of the text is determined through an instructional design process (e.g. Reigeluth, 1983), implemented when the instruction is developed, and refined during the formative evaluation stage. The manipulation of text density, however, is typically done after the materials have been developed by removing redundant and nonessential information. The following example from a statistics book illustrates the manipulation of the text density in a paragraph. The top excerpt is the standard, high-density format; the bottom one is a low density variation.

> The second-most used measure of central tendency is the median; its definition is straightforward. The median was introduced several times in Chapter 3, as in the median family income and median (*P* v *50*) IQ. The *median is the 50th percentile of a distribution*—the point below which half the observations fall. In any distribution, there will always be an equal number of cases above and below the median. The interpretation of the median is even more direct and clear-cut than that of the mean (Hopkins & Glass, 1978, p. 53).

> The median: 2nd most popular central tendency measure. Chapter 3 examples: median family income and median (*P* v *50*) IQ. In 50th percentile of any distribution: half the scores fall below it; half above. Easier to interpret than mean.

Both of the paragraphs convey the same basic ideas, but the details and nonessential words are removed. The low-density version has approximately 50% fewer words and would require less screen space for the display, thus allowing more screen area to implement spatial cues. The first hypothesis for this study was that the low-density narrative would promote better learning and more favorable attitudes on CBI lessons by reducing reading and cognitive processing demands of the screen displays.

Applying these rules to a conventional textbook chapter of 2,123 words resulted in a low-density version of 1,189 words (Morrison, Ross, & O'Dell, 1988; Ross, Morrison, and O'Dell, 1988) (see Figure 12.1). The decrease in the number of words in the low-density text allows the designer to make liberal use of white space, and vertical typography to highlight and to group ideas while maintaining an appropriate level of contextual support on individual screens.

The primary focus of the text-density studies was to evaluate the effectiveness of the low-density material for learning. We predicted that low-density CBI narrative would promote better learning and more favorable attitudes by reducing reading and cognitive demands of the screen displays. A second interest was the effect of text density as a learner control (LC) variable. Prior research on learner control has typically allowed the learner to determine the amount of content included in the lesson. Recent results in which learners were allowed to determine the amount of content they received have been negative (Carrier, Davidson, & Williams, 1985; Fisher, Blackwell, Garcia, & Greene, 1975; Ross & Rakow, 1981; Tennyson, 1980). Text density, on the other hand, allows the learner to manipulate the 'contextual' properties of the lesson that affects how the lesson appears rather than changing the basic information content. Learner control choices then, were assumed to be less dependent relative to other LC variables on prior knowledge or skill in the subject area.

To investigate these questions concerning density variations and learner control, we conducted two text density studies (Morrison et al., 1988; Ross et al., 1988). In the first study, 48 subjects received either a print or CBI statistics lesson of either high-density, low-density, or learner control of text materials in

The median corresponds to the middle frequency score in a ranked set of data

Half the scores will be higher
Half will be lower

	X	f
	Hi	50%
Median	------------	------------
	Lo	50%

If N=40 (40 scores), median = 20th score
If N=17, median = 8.5 highest score

Median corresponds to the 50th percentile

Higher than half the scores
Lower than half

Figure 12.1. High and low density text

a print or CBI format. Stimulus materials consisted of a statistics lesson adapted from Ross (1983). The high-density version was patterned after the original text and was 18 pages long with 2,123 words. The low-density version was developed from the high density version by (a) reducing the sentences to their main ideas, (b) using at outline form, (c) deleting sentences that summarized or amplified without adding additional information, and (d) presenting the information in 'frames' of limited information similar to programmed instruction. The low-density version consisted of 1,189 words, a 56% savings, and was 17 pages long. A similar reduction in page length was not realized due to the liberal use of white space in the low-density version. Computer versions of the stimulus materials were designed independently to arrive at what we determined to be acceptable screen designs. The low density CBI lesson consisted of 49 frames and the high-density CBI lesson consisted of 66 frames. Enhanced Apple IIe's with monochrome monitors were used for the CBI lessons which were presented on the graphic screen with a font similar to the standard 40 column text font. The same stimulus materials were used in the second study, however, an additional treatment was added. Subjects (n = 221) were allowed to select either a print or CBI presentation for the lesson.

In both studies, subjects in the high-density and low-density treatments scored equally well on the achievement posttest. While there was no significant difference in achievement scores between the two density conditions, the high-density group took significantly longer (34%) to complete the lesson. In addition, learner control subjects selected the low-density materials approximately 65% of the time in Study II. A comparison of responses between subjects who received the low-density material in print mode and those who received it in CBI mode, found that the CBI subjects rated the materials as more sufficient. These results suggest that low-density text is a viable alternative for CBI screen design.

Implications for screen design

Low-density text provides the instructional designer with an alternative method for displaying text originally designed for a print format. This strategy also provides a means for compensating for the slower reading rate often associated with CBI displays (Hansen, Doring, & Whitlock, 1978; Muter, Latremouille, & Treurniet, 1982) by presenting fewer words without a corresponding decrease in comprehension.

Another finding of interest from the two studies was the learner control results. The results indicated that less skilled readers were more likely to select high-density text which offered more contextual support. Better readers were more likely to select low-density text which provided adequate contextual support and reduced their reading time. Learner control subjects also varied their selection of high- and low-density text which suggests they were attempting to adjust their strategy as their learning needs varied. These results suggest that allowing subjects to vary the contextual properties of the lesson may be a potentially effective form of learner control.

Screen density as a design variable

Prior research on typographical variables and content manipulation have provided useful guidelines for screen design; however, they have not addressed the issue of how much information an expository frame should contain. For example, the International Reading Association Computer and Technology Reading Committee (1984) recommends using 'clear and legible' displays with 'appropriate margins and interline spacing', but provides no operational guidelines or specifications to define these qualities. To provide designers with clearer recommendations for optimum density levels, the screen density construct must be operationalized and precisely defined. Tullis (1983, 1988), for example, provides a method for calculating screen density by determining the percentage of screen spaces that contain a character or are adjacent to a character.

Studies of the effect of screen density, which come primarily from human factors research, have found that performance error rates increase as the density of a display increases (Burns, 1979; Coffey, 1961; Mackworth, 1976; Ringel and Hammer, 1964). Research, however, on the upper limit of screen density has yielded disparate recommendations ranging from 15% (Danchak, 1976) to 31.2% (Smith, 1980, 1981, 1982) all the way to 60% (NASA, 1980). Our first interest in the screen density studies was to determine which level of screen density subjects preferred when viewing a fixed amount of material.

A second interest was the possible influence of the *type* of material presented on how different screen designs were viewed. For example, Grabinger's (1983) results suggested that learners prefer a low-density screen when judging a

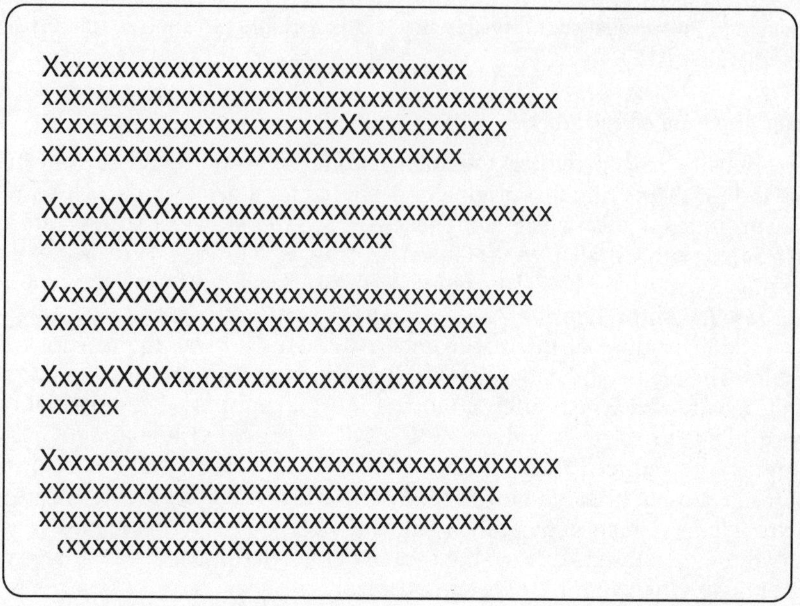

Xxxxxxxxxxxxxxxxxxxxxxxxxxxxxxxx
xxxxxxxxxxxxxxxxxxxxxxxxxxxxxxxxxxxxxx
xxxxxxxxxxxxxxxxxxxxxxxXxxxxxxxxxxx
xxxxxxxxxxxxxxxxxxxxxxxxxxxxxxxx

XxxxXXXXxxxxxxxxxxxxxxxxxxxxxxxxxxxxx
xxxxxxxxxxxxxxxxxxxxxxxxxxx

XxxxXXXXXXxxxxxxxxxxxxxxxxxxxxxxxxxx
xxxxxxxxxxxxxxxxxxxxxxxxxxxxxxxxxx

XxxxXXXXxxxxxxxxxxxxxxxxxxxxxxxxxx
xxxxxx

Xxxxxxxxxxxxxxxxxxxxxxxxxxxxxxxxxxxxxxx
xxxxxxxxxxxxxxxxxxxxxxxxxxxxxxxxxxx
xxxxxxxxxxxxxxxxxxxxxxxxxxxxxxxxxxxx
xxxxxxxxxxxxxxxxxxxxxxxxx

Figure 12.2. Sample screen using Twyman's notation.

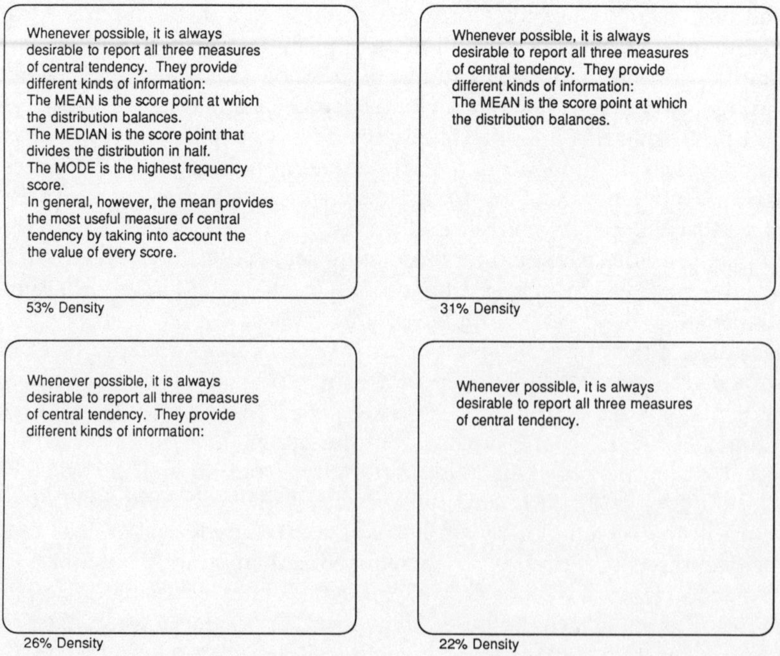

Whenever possible, it is always desirable to report all three measures of central tendency. They provide different kinds of information:
The MEAN is the score point at which the distribution balances.
The MEDIAN is the score point that divides the distribution in half.
The MODE is the highest frequency score.
In general, however, the mean provides the most useful measure of central tendency by taking into account the the value of every score.

53% Density

Whenever possible, it is always desirable to report all three measures of central tendency. They provide different kinds of information:
The MEAN is the score point at which the distribution balances.

31% Density

Whenever possible, it is always desirable to report all three measures of central tendency. They provide different kinds of information:

26% Density

Whenever possible, it is always desirable to report all three measures of central tendency.

22% Density

Figure 12.3. Screen density variations.

screen using Twyman's (1981) notation of substituting x's and o's for the letters (see Figure 12.2). In contrast, judgments of realistic materials would appear to demand greater awareness of and reliance on contextual properties (e.g., proximal supporting text) that help to increase the meaning of the information being read. Thus, it is not clear that preferences for low-density screens similarly apply to realistic lesson materials, especially since the low-density designs present the material in smaller thought units and consequently also necessitate an increased number of lesson frames. We predicted that when viewing realistic displays learners would also prefer low density materials. A third research interest was the preferences of users differing in degree of CBI experience, namely graduate instructional design students versus undergraduate education students (Ross et al., 1988).

The two studies on screen density used one screen of information (selected from the text density stimulus materials) presented in four different formats. The first format consisted of one frame with all the content. The second format divided the context across two frames, and the third and fourth formats divided the content across three and four frames respectively (see Figure 12.3). The divisions in the frames were made at logical points in the narrative. Screen density was then calculated by counting all spaces and spaces contiguous to a character and then dividing by the total number of characters possible on the display. An average density was calculated for the multiple screen displays. The resulting density levels for the screens ranged from 50% for the single frame, 31% for two frames, 26% for three frames, to 22% for four frames. Subjects in the third study viewed each frame of the four designs. In contrast to recommendations in the literature (Allessi & Trollip, 1985; Bork, 1984, 1987; Grabinger, 1983; Heines, 1984; Hooper & Hannafin, 1986) for designing lower density screens, the results of the study showed that subjects tended to prefer higher-density screens. The relative strong preference for the 31% density screen may suggest the subjects were trying to balance aesthetic properties with either or both the degree of (a) contextual support or (b) number of frames in the lesson. If this preference was due to the number of screens, it would seem likely then, the preferences for lower-density designs would increase if subjects view *only* the first screen of each density level.

In the fourth study, subjects viewed only the first frame of the four formats. Although there was not a significant difference between preferences for the different levels, the subjects preferred the higher density designs over the lower density designs. The results from these two studies suggest that subjects prefer adequate contextual support as provided by either the single frame design or the two screen design as opposed to screens with large amounts of white space.

Implications for screen design

The finding that subjects prefer high-density screens when using realistic materials is in contrast to the recommendations in the literature which suggest the use of low-density screens with adequate white space and vertical typography. Thus, instructional designers should give careful consideration to screen design to ensure that adequate contextual support is provided on each screen even at the expense of white space that has been added for aesthetics.

The final study extended the two studies on screen density to determine subject preferences when using different types of stimulus material. An issue of concern in the two screen density studies was the external validity of the presentation format (e.g., full text or only the first screen) and of the stimulus materials. Three different types of stimulus materials were manipulated. One type was standard English. Second was a nonsense material developed according to Weaver's (1948) guidelines (see Figure 12.4). This nonsense material (called third-order approximation to English) has the same structure as English and presents a similar pattern in print or on the screen giving the reader the same visual image as when reading realistic text. A third type of stimulus material was developed using the guidelines proposed by Twyman (1981). Screens were designed for the Twyman and nonsense materials with 50%, 31%, 26% and 22% density levels based on the standard English versions used in the previous studies. The study was a 2 (format: full text, first screen only) x 3 (stimulus type: English, Approximation to English, Twyman) paired-comparison design. We predicted that subjects would indicate a preference for high density screen designs when viewing realistic screens and for lower density

Res hadged infivel slesahn fifo
nty histrespl stramas hivlarn ngelmosy
oowhe brkichigke wa. Allalid ofle
whr oolt ho idr igonkest haosc:

Blm cupagonestale ne onot thi hethec
whotuedec nechene dthedis.

Iscas me ggidag ounoaw hoit arlnet
beongene rsit anoittw igu mapumi.

Wes hothind tns pey alyintoll yyphi.
n hie

Dulig wrimond onthalm folere dtststhom
iliserin drdg velo tenger ee owatt
hsintchema wiffrougnd paby ooherecg
maind ondalltwin wyovede.

Figure 12.4. Third-order approximation to English.

screen designs when viewing content free screens (approximation to English and Twyman). We also predicted that the preferences would remain constant regardless of the presentation format.

The results indeed showed a significant tendency for subjects to prefer higher-density screens for the realistic lesson material, but lower-density screens for the artificial (approximation to English and nonsense) text forms. Further, this pattern was more pronounced under the multiple-screen than single-screen condition. These results suggest that aesthetic appeal (as represented by 'spacier' screens) becomes secondary to textual support (higher content-density screens) when there is a need to comprehend the material displayed. An underlying issue concerns the questionable external validity of studies that use artificial stimuli to identify effective screen designs. That is, in the present case, selecting the more visually attractive low-density screens neither impacted on comprehension when artificial text was employed nor on lesson length (the total frame count) when only one screen was affected. With realistic lesson material, however, high density screens worked in the opposite manner to increase potential comprehensibility and reduce lesson length.

Discussion

Our studies on text density and screen density suggest two additional variables to consider when designing CBI text screens. First, low-density text format is a viable alternative to the standard text format used in printed materials. The text-density algorithm (Morrison et al., 1988) can be used to convert lengthy text materials into an acceptable format for CBI screen design when the system requires the use of such materials. Implementation of this strategy typically reduces the amount of text by 50% while also significantly reducing reading time.

Second, in contrast to previous studies and recommendations in the instructional design literature (Allessi & Trollip, 1985; Bork, 1984, 1987; Grabinger, 1983; Heines, 1984; Hooper & Hannafin, 1986), subjects indicated a strong preference for learning from high density screens as opposed to low-density screens. These high-density screens also provided full contextual support for the main ideas. When the information was divided over two, three, or four frames, the contextual support was also reduced. Thus, subjects had to read more frames to obtain the same information. These results suggest that the use of realistic stimulus materials may produce different results from those obtained with nonrealistic stimulus materials (Grabinger, 1983) or with informational (e.g., machine status) displays (e.g., Danchak, 1976; Smith, 1980, 1981, 1982).

Future research on CBI screen designs should investigate the use of text density and varying screen density with different content areas and tasks with different processing demands. Specifically, the expository presentations used in the present research for explaining the rationale for the operation of different statistical indices seemed to require the high contextual support provided by high-density screens. Other types of learning, such as memorizing foreign language vocabulary sentences, may be impaired by lower-density contexts. The extension of these investigations of text and screen properties to applications involving information used primarily for review rather than instruction (e.g., on-line help screens) should also prove valuable for improving CBI designs.

References

Allessi, S. M. & Trollip, S. R. (1985). *Computer-based instruction: Methods and development*. Englewood Cliffs, NJ: Prentice-Hall, Inc.

Basset, J. H. (1985). *A comparison of the comprehension of chunked and unchunked text presented in two modes: Computer and printed page*. Unpublished doctoral dissertation, Memphis State University.

Bork, A. (1984). *Personal Computers for Education*. Cambridge, MA: Harper & Row.

Bork, A. (1987). *Learning with personal computers*. New York: Harper & Row, Publishers, Inc.

Burns, D. (1979). A dual-task analysis of detection accuracy for the case of high target-distractor similarity: Further evidence for independent processing. *Perception and Psychophysics*, 25, 185–196.

Carrier, C. A., Davidson, G., & Williams, M. (1985). The selection of instructional options in a computer-based coordinate concept lesson. *Educational Communications and Technology Journal*, 33, 199–212.

Carver, R. P. (1970). Effect of 'chunked' typography on reading rate and comprehension. *Journal of Applied Psychology*, 54, 288–296.

Coffey, J. L. (1961). A comparison of vertical and horizontal arrangements of alpha-numeric material—Experiment 1. *Human Factors*, 3, 93–98.

Danchak, M. M. (1976). CRT displays for power plants. *Instrumentation Technology*, 23, 29–36.

Davidson, A., & Kantor, R. (1982). On the failure of readability formulas to define readable texts: A case study from adaptations. *Reading Research Quarterly*, 17, 187–209.

DiVesta, F. J. & Finke, F. M. (1985). Metacognition, elaboration, and knowledge acquisition: Implications for instructional design. *Educational Communications and Technology Journal*, 33, 285–293.

Duchastel, P. C. (1983). The use of summaries in studying texts. *Educational Technology*, 23, 36–41.

Feibel, W. (1984). Natural phrasing in the delivery of text on computer screens: Discussion of results and research approaches. In Bonnett, D. T. (Ed.), *Proceedings of the Sixth Annual National Computing Conference*. Dayton, OH.

Fisher, M. D., Blackwell, L. R., Garcia, A. B., & Greene, J. C. (1975). Effects of student control and choice on engagement in a CAI arithmetic task in low-income school. *Journal of Educational Psychology*, 67, 776–783.

Frase, L. T., & Schwartz, B. J. (1971). Typographical cues that facilitate comprehension. *Journal of Educational Psychology*, 71, 197–206.

Frayer, D. A., Frederick, W. C., & Klausmeir, H. J. (1969). *A schema for testing the level of concept mastery*. (Working paper No. 16). Madison: Wisconsin Research and Development Center for Cognitive Learning.

Gerrel, H. R., & Mason, G. E. (1983). Computer-chunked and traditional text. *Reading World*, 22, 241–246.

Grabinger, R. S. (1983). *CRT text design: Psychological attributes underlying the evaluation of models of CRT text displays.* Unpublished doctoral dissertation, Indiana University.

Hansen, W. J., Doring, R. R., & Whitlock, L. R. (1978). Why an examination was slower online than paper. *International Journal of Man-Machine Studies*, 10, 507–519.

Hartley, J. (1980). Spatial cues in text: Some comments on the paper by Frase and Schwartz (1979). *Visible Language*, 24, 62–79.

Hartley, J. (1985). *Designing Instructional Text*. (2nd ed.) London: Kogan Page Ltd.

Hartley, J. (1987). Designing electronic text: The role of print-based research. *Educational Communication and Technology Journal*, 35, 3–17.

Heines, J. M. (1984). *Screen design strategies for computer-assisted instruction*. Bedford, MA: Digital Press.

Hooper, S., & Hannafin, M. J. (1986). Variables affecting the legibility of computer generated text. *Journal of Instructional Development*, 9, 22–29.

Hopkins, K. D., & Glass, G. V. (1978). *Basic statistics for the behavioral sciences*. Englewood Cliffs, NJ: Prentice-Hall.

International Reading Association Computer Technology and Reading Committee (1984). Guidelines for educators on using computers in schools. *Reading Research Quarterly*, 20, 120–122.

Keenan, S. (1984). Effects of chunking and line length on reading efficiency. *Visible Language*, 28, 61–80.

Lancaster, F. W., & Warner, A. (1985). Electronic publication and its impact on the presentation of information. In Jonassen, D. H. (Ed.), *The technology of text: Principles for structuring, designing, and displaying text* (Vol. 2). Englewood Cliffs, NJ: Educational Technology Publications.

Mackworth, N. H. (1976). Stimulus density limits the useful field of view. In Monty, R. A. and Senders, J. W. (Eds.) *Eye movements and psychological processes*. Hillsdale, NJ: Erlbaum.

Morrison, G. R., Ross, S. M., & O'Dell, J. K. (1988). Text density level as a design variable in instructional displays. *Educational Communication and Technology Journal*, 36, 103–115.

Muter, P., Latremouille, S. A., & Treurniet, W. C. (1982). Extending reading of continuous text on television screens. *Human Factors*, 24, 501–508.

NASA. (1980). *Spacelab display design and command usage guidelines*. (Report MSFC-PRC-711A). Huntsville, Al: George Marshall Space Flight Center.

O'Shea, L. T., & Sinclair, P. T. (1983). The effects of segmenting written discourse on the reading comprehension of low- and high-performance readers. *Reading Research Quarterly*, 18, 458–465.

Reder, L. M., & Anderson, J. R. (1980). A comparison of texts and their summaries: Memorial consequences. *Journal of Verbal Learning and Verbal Behavior*, 19, 121–134.

Reder, L. M., & Anderson, J. R. (1982). Effects of spacing and embellishments on memory for the main points of a text. *Memory and Cognition*, 14, 64–78.

Reder, L. M., & Charney, D. H., & Morgan, K. I. (1986). The role of elaborations in learning a skill from the main points of a text. *Memory and Cognition*, 10, 97–102.

Reigeluth, C. M. (1983). *Instructional-design theories and models: An overview of their current status*. Hillsdale, NJ: Lawrence Erlbaum Associates, Publishers.

Richardson, J. J. (1980, October). *The limits of frame-based CAI*. Paper presented at the annual conference of the Association for the Development of Computer-Based Instructional Systems, Atlanta, GA.

Ringel, S., & Hammer, C. (1964). Information assimilation from alphanumeric displays: Amount and density of information presented (Tech. Report TRN141). Washington, DC US Army Personnel Research Office (NTIS No. AD 6021 973).

Ross, S. M. (1983). *Introductory Statistics*. Danville, IL: Interstate.

Ross, S. M., Morrison, G. R., & O'Dell, J. K. (1988). Obtaining more out of less text in CBI: Effects of varied text density levels as a function of learner characteristics and control strategy. *Educational Communication and Technology Journal*, 36, 131–142.

Ross, S. M., & Rakow, E. A. (1981). Learner control versus program control as adaptive strategies for selection of instructional support on math rules. *Journal of Educational Psychology*, 73, 745–753.

Sherrard, C. (1988). What is a summary? *Educational Technology*, 28, 47–50.

Smith, S. L. (1980). *Requirements definition and design for the man-machine interface in C^3 system acquisition*. (Technical Report ESD-TR-80-122). Bedford, MA: USAF Electronic Systems Division. (NTIS No. AD A087 528).

Smith, S. L. (1981). *Man-machine interface (MMI) requirements definition and guidelines: A progress report*. (Technical Report ESD-TR-81-113). Bedford, MA: USAF Electronic Systems Division. (NTIS No. AD A096 705).

Smith, S. L. (1982). *User-system interface design for computer-based information systems*. (Technical Report ESD-TR-82-132). Bedford, MA: USAF Electronic Systems Division. (NTIS No. AD A115 853).

Tennyson, R. D. (1980). Instructional control strategies and content structures as design variables in concept acquisition using computer-based instruction. *Journal of Computer-Based Instruction*, 3, 84–90.

Tullis, T. S. (1983). The formatting of alphanumeric displays: A review and analysis. *Human Factors*, 25, 657–682.

Tullis, T. S. (1988). Screen Design. In Helander, M. (Ed.), *Handbook of Human-Computer Interaction*. (pp. 377–410). North-Holland: Elsevier Science Publishers B. V.

Twyman, M. (1981). Typography without words. *Visible Language*, 15, 5–12.

Weaver, W. (1949). Recent contributions to the mathematical theory of communication. In Shannon, C. E. & Weaver, W. (Eds.), *The mathematical theory of communication*. (pp. 1–28). Urbana: The University of Illinois Press.

Suggested further reading

Hansen, W. J. & Haas, C. (1988). Reading and writing with computers: a framework for explaining differences in performance. *Communications of the A. C. M.* , 31, 9, 1080–1089.

Hartley, J. (1987). Designing electronic text: the role of print-based research. *Educational Communication & Technology Journal*, 35, 1, 3–17.

Lachman, R. (1989). Comprehension aids for on-line reading of expository text. *Human Factors*, 31, 1, 1–15.

Tullis, T. S. (1988). Screen design. In Helander, M. (Ed.) *Handbook of Human-Computer Interaction*. Amsterdam: Elsevier.

Wright, P. (1988). Issues of content and presentation in document design. In Helander, M. (Ed.) *Handbook of Human-Computer Interaction*. Amsterdam: Elsevier.

Chapter 13

Using Style Sheets, Templates and the Features of Publishing Software to Facilitate the Development of Printed Study Materials*

David Kember

Introduction

This chapter is concerned with the ways in which publishing software can assist in the development and production of print-ready master copies for printed study materials. The term study materials is taken as embracing printed matter produced for courses under banners such as distance education, open learning, continuing education, self-study and independent learning.

Characteristics of study materials

All textbooks contain a variety of features or components. The text will be divided into paragraphs. There are normally headings, often of several levels. Many textbooks have diagrams which often have associated captions.

Study materials need these general features and others as well. Gagné and Briggs (1979, p. 157), who have influenced the instructional design of many such courses, divide instruction into nine types of event. Rowntree (1986, p. 163) lists the following sixteen devices to aid access to self-instructional material:

Before

- Explanatory title
- Contents list
- Concept map/Flow diagram
- List of objectives
- Pre-test

---

* *This article first appeared in* Educational Training and Technology International, *1989, 26, 72-78. It is reprinted here with kind permission of the author and the Association of Educational Training and Technology.* Educational Training and Technology *is published by Kogan Page Ltd. on behalf of AETT, BMA House, Tavistock Square, London WC1H 9JP, England.*

During

- Introduction
- Links with other lessons
- Headings
- Numbering systems
- Instructions
- Verbal signposts
- Visual signposts
- Summaries

After

- Glossary
- Post-test
- Index

For the purpose of this article these access devices and other more general features found in textbooks will be referred to as components.

The desirability of a consistent format

Publishers normally strive for consistency in the appearance or format of components within a book or journal. There is no doubt that some of this striving arises from the sense of orderliness and neatness which seems to characterize all editors. Fortunately the characteristic benefits the reader because consistency of format aids access. For example, if a sub-heading in a book is to be identified as a sub-heading it should have the same appearance as other headings of the same type. The textual design should ensure that headings stand out clearly from the main body of text. A sub-heading should be distinct from a main heading, but ideally the design should have an element of similarity that characterizes the different levels of heading as a group of components.

As study materials generally have more types of component than ordinary books, the need for clear distinction between components becomes more important. Access is particularly important in self-study materials as it is recognized that students can legitimately use them in a wide variety of ways and sequences. In a diary survey of just 25 distance education students, Clyde et al. (1983) found four distinct patterns for studying modules of distance education courses. The authors stressed that there were many variations on the four main patterns.

There is a clear cut case for consistency in the format for components within the booklets for an individual subject. It can also aid student access if there is a consistency of format between the subjects which make up a course or programme. However, the format specifications should be flexible enough for essential instructional design differences. An introductory subject would normally be quite distinct, in instructional design, from a final year project. The textual design should be sensitive to these differences and also be able to enhance salient instructional features.

The case for consistency of format between courses is concerned less with student access and more with institutional staff development and the desirability of a house style. The greater the consistency between specified formats for courses, the easier it is for operators and editors to accurately and consistently reproduce the specified format. However, when a house style is applied across a diverse range of courses, it is even more important that the format permits flexibility for the instructional design of subject areas and individual units.

Methods for ensuring that a format is consistent

Traditional methods

Traditionally, publishing houses have used documents with titles such as 'style manual', 'mark- up manual' or 'guide to contributors', to define a house style. These documents contain instructions for the format of normally occurring components.

As with other 'bibles' the mere existence of the instruction is insufficient to ensure consistent adherence. Those entering the text are normally given some form of instruction in following the defined formats. Monitors, usually known as editors, are needed to interpret the rules and check to see that they are correctly and consistently applied.

In addition to style manuals there are invariably unwritten conventions. Keyboard operators are taught conventions or acquire habits. Editors retain practices from previous experiences or establish interpretations for areas which are not covered by the style manual. Some unwritten conventions become common to an organization while others remain characteristic of an individual.

Use of computers

As with many other functions, the computer can be used to facilitate the application of standard formats to components in study materials. Timmers (1986) has described the use of course authoring templates on microcomputers.

The templates described by Timmers were derived from mark-up guides. Elements common to courses were entered in the designated position and typeface on the word processor. The majority of elements, which are not common to courses, have to be indicated by Xs for text and Os for numbers.

The value of such templates is clear for pages which usually appear at the beginning or end of courses or units, such as cover, title, contents and copyright pages. These pages often have common phrases between courses and unique information appears in standard positions. However, the application must be of limited value in the bulk of a course where there is no fixed order for components such as headings, paragraphs of text, figure captions and questions.

Style sheet

With publishing/word processing software now available it is possible to use computers much more extensively to apply consistent formats to components of study materials than the templates described by Timmers (1986). This statement records the advances in software which would presumably not have been available to Timmers.

The ideas on using computers for study material development which are described in this chapter, are based on the author's experiences of implementing new computer systems at two institutions. At Capricornia Institute, Australia, a token ring network of Apollo Domain computers was installed with Interleaf WPS publishing software. At the Hong Kong Polytechnic, Microsoft Word was the main item of software used on Macintosh microcomputers.

Both items of software feature methods of applying consistent formats to items of text which are identified as being the same type of component. As Microsoft Word is much more common than Interleaf WPS, the Macintosh implementation will be described and Microsoft Word terminology will be used in this chapter.

In Microsoft Word the facility for applying consistent formats to components is known as a style sheet. A style or format can be defined for each type of component which can include:

- font style
- font size
- ruler
- tab stops
- inter-line spacing
- space before and after component
- justification or alignment
- borders
- position relative to other components

The text can then be treated as a series of components separated by returns. Each component is tagged with the correct style by highlighting it and selecting the appropriate style from a menu. Alternatively a few keystrokes can specify a style. The same defined format will then be applied to all components tagged with the same style label. If changes are made to the format defined for that style, these can be applied automatically to all components with the same style label.

For example, a main heading might be defined as Helvetica 14 pt bold, left aligned, with 28 pt of space above and 20 pt below. If the style definition is changed to italic rather than bold then all phrases throughout the document, tagged as main headings, will almost instantly change to italics.

Microsoft Word was chosen in preference to other software, most importantly because it has this facility. With other Macintosh word processing and/or desktop publishing software, at the time of the implementation the format for each component had to be remembered and applied each time a component appeared. Since this time Page Maker has added a style sheet facility.

In addition, Microsoft Word has been developed to the extent that it now has the properties of both word processing and desktop publishing software. It can incorporate graphics from various graphics packages and has other features found in desktop publishing software. It is therefore possible to use Microsoft Word rather than a combination of a word processing package and desktop publishing software. Other useful features of Microsoft Word will be described later.

Using templates with a style sheet

A standard template has been prepared for distance education courses developed at the Hong Kong Polytechnic. Subjects within courses are divided into units, which usually correspond to one week's work.

At first sight the template document appears to contain very little. Visible on screen are a few headings, some page breaks and a few phrases of text which are repeated from unit to unit. These elements all belong to the first and last few pages of a unit which are standardized. A typical sequence for the first four pages of a unit might be title page, list of unit contents, page containing an advance organizer and objectives, and then a page devoted to a recurring concept map with relevant concepts highlighted. Following the body of the unit might be a summary or glossary and review questions.

There needs to be variation between subjects on these elements according to the instructional design of the course. For example, some courses might be better suited to objectives at the end rather than the beginning. It is very easy to modify the template to suit individual subjects. Once the template is set up for a subject it helps to ensure that each unit has a consistent format for the first and last few pages.

The most important feature of the template is revealed if the style menu is examined. On the style sheet 25 styles are stored for components which commonly appear in study materials.

The template can be used in three ways:

- as a paper version
- for keying-in text directly
- for transferring text files from other computers and software.

These three ways of using the template are discussed below.

Paper version

The Hong Kong Polytechnic has recently started developing a number of distance education courses with quite large teams of writers, most of whom have no experience of developing printed study materials. A printed version of the template serves as a guideline to these authors on the content and format of a unit. For these purposes the screen version of the template is supplemented with brief notes on the form and function of the items specified by the template. The course team is first introduced to the template during a workshop on writing study materials. It continues to serve as a guide throughout the writing process.

Keying-in text to the template

Units are prepared by the authors either on disc or as written manuscripts. When a written version is received, the instructional designer works on the unit and marks alongside the text the name of the style to be applied to each component. The operators then key-in the text under the pre-set headings in the template. The marked styles are applied as the keying-in takes place.

Transferring text files into the template

Nowadays many course writers have access to computers with word processing software, either at work or in the home. An increasing number of academics prefer to word process rather than hand write documents. When establishing computer systems for the production of study materials it is important that they can accept input from potential authors' hardware and software. Otherwise extensive re-keying and proofreading must be undertaken.

Attempting to standardize hardware and software is doomed to failure. It is difficult to persuade the departments of an institution that one type of computer serves all their needs. It is well nigh impossible to persuade individuals to buy the same brand for use in their homes. A better strategy is to develop routines for transferring text files into the production system.

Once a text file has been transferred to the Macintosh it can be 'cut and pasted' into appropriate positions in the template. Formatting is then applied by going through the document applying the selected style to each component.

Other software features used in the template

Page set-up

The standard page specifications are based on a modular design (Hartley, 1985, p.18) with a wide left margin reserved for marginal keywords and icons. The main body of text then averages roughly 10–12 words per line which makes for optimum readability (Trevitt, 1980).

The page set-up dialogue box also allows the maximum page length to be set. Facing page options and gutter margins can also be specified.

Side by side paragraphs

An unusual feature of Microsoft Word is the option of specifying that paragraphs appear side by side. This feature is used to position marginal keywords and icons alongside relevant paragraphs in the main body of the text.

Running headers and footers

The template specifies running headers with the unit title. The operator has only to edit in the unit title. The page number appears as a footer. Both header and footer appear on the outside of the page as the facing page option is selected. Headers and footers both have a specified hairline border as a design feature.

Automatic table of contents and indexing

Microsoft Word has a facility for automatically compiling a table of contents from hidden identifiers. These identifiers are placed by headings in the template and are added alongside headings as these are keyed-in with the text. The software will then compile a table of contents with page numbers and format it as specified by the style, for each level of entry, in a table of contents. An overall table of contents is used at the start of a course and a unit table of contents at the start of each unit.

The indexing facility works on a similar principle; compiling an index from hidden identifiers alongside words or phrases in the text. The facility has not yet been used extensively as trial versions of courses are produced in short sections. The index facility will be evaluated fully when a complete subject has been compiled on the Macintosh and is ready for reprinting. Indexes are useful access devices which rarely appear in study materials.

Graphics input

The modular page design with a wide left margin permits the use of icons to highlight components. Standard icons have been prepared to highlight important instructions and to identify different types of questions and activities, readings from textbooks or readers, and recommended readings. These icons are stored in a document which can be placed in a window alongside the text. Icons can then be copied and pasted into desired positions alongside the text. Figure 13.1 shows the appearance of a standard size Macintosh screen with part of the icon document alongside a window containing the template document. The arrangement is acceptable with standard size screens, but larger screens are more convenient if extensive graphics work is undertaken.

Graphics, drawn with the various graphics packages available, are also pasted into the text file where minor scaling to fit available space can be

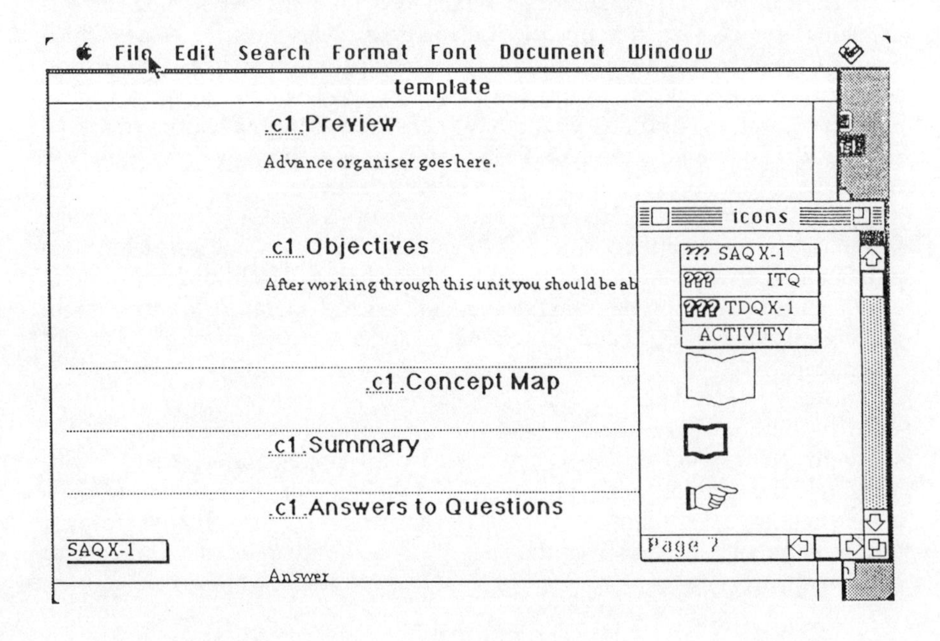

Figure 13.1. Part of the icon document with a template window.

performed. Art work can also be scanned in. Scanning eliminates paste-up of artwork but is not ideal for courses with numerous illustrations. Scanned images considerably slow down both screen manipulation and printing. The images also have extensive memory requirements.

Complex scientific or mathematical equations and formulae are also pasted into Microsoft Word documents. The formulae are first generated with Expressionist.

Output

Documents produced using the template are normally printed out with an Apple Laser Writer Plus. The output is of 'typeset appearance' but at lower resolution than that produced by a typesetter. Experiments have recently been performed with outputting documents on a Linotronic 300 typesetter located elsewhere in the Polytechnic. No alterations, coding or re-keying are required. The document is output with the same format and appearance as on a laser printer but at considerably higher resolution and contrast.

Operator implementation

Even from the brief description in this chapter, it will be obvious that Microsoft Word offers both facilities and precision of text control not found in most word processing or indeed some 'desktop publishing' software. The trade-off is that it is more complex to operate if all facilities are utilized.

The complexity of software must be considered in respect to the potential user. For the two implementations which provided the experience for this chapter, there were existing operators using dedicated word processors which were replaced by the new computer systems. In both instances the existing operators were expected to use the new system. The training of a typical operator at both institutions would have consisted of typing lessons at school and a short course in using a dedicated word processor. Any other expertise was picked up on the job, usually from other operators.

It is quite unreasonable to expect such operators to possess design and page layout skills. Few can be expected to master the full range of facilities in publishing software. At the Hong Kong Polytechnic it would have been quite unrealistic to expect typographic design skills as the operators are paid about one-eighth of the salary of an instructional designer and are working in a second language.

The use of a template and style sheet allows operators to use publishing type software without either design skills or a comprehensive knowledge of the software. The textual design is provided by the implementer(s) and standardized in the template and style sheet. The implementers therefore need skills in instructional text design and a thorough knowledge of the software. To enter text and apply styles, however, the operators need only to learn a limited part of the software.

References

Clyde, A., Crowther, M., Patching, W., Putt, I. and Store, R. (1983). How students use distance teaching materials: an institutional case study. *Distance Education*, 4, 1, 4–26.

Gagné, R. M. and Briggs, L. J. (1979). *Principles of Instructional Design*. New York: Holt, Rinehart and Winston.

Hartley, J. (1985). *Designing Instructional Text* (2nd edition). London: Kogan Page.

Rowntree, D. (1986). *Teaching Through Self-instruction: a Practical Handbook for Course Developers*. London: Kogan Page.

Timmers, S. (1986). Microcomputers in course development. *Programmed Learning and Educational Technology*, 23, 1, 15–23.

Trevitt, J. (1980). *Book Design*. Cambridge: Cambridge University Press.

Suggested further reading

Hartley, J. (1986). Planning the typographical structure of instructional text. *Educational Psychologist*, 21, 4, 315–332.

Lefrere, P. (1989). Design aids for constructing and editing tables. *British Library Research Paper No. 61*. London: British Library.

Miles, J. (1987). *Design for Desktop Publishing*. London: Gordon Fraser.

Sylla, C., Drury, C. G., & Babu, A. J. G. (1988). A human factors design investigation of a computerised layout system of text-graphic technical materials. *Human Factors*, 30, 3, 347–358.

Chapter 14

Authoring and Editing Hypertext*

Ben Shneiderman

Authoring or editing hypertext

For at least the last three thousand years authors and editors have explored ways to structure knowledge to suit the linear print medium. When appropriate, authors have developed strategies for linking related fragments of text and graphics even in the linear format. Now, hypertext encourages the non-linear interconnecting links among nodes.

Restructuring knowledge to suit this new medium is a fascinating experience (Koved, 1985; Kreitzberg and Shneiderman, 1988; Weldon, Mills, Koved and Shneiderman, 1985). The first challenge is to structure the knowledge in a way that an overview can be presented to the reader in the root document or introductory article. The overview should identify the key subsidiary ideas and the breadth of coverage. Paper books present a clear vision of their boundaries so readers can know when they have read it all, but in the hypertext world other mechanisms must be created to give the reader a sense of scope and closure. The overall structure of articles must make sense to readers so that they can form a mental image of the topics covered. This facilitates traversal and reduces disorientation. Just as important is the reader's understanding of what is not in the database. It can be terribly frustrating if readers think that something of interest is in the database, but they can neither find it nor convince themselves that it is not there.

In writing articles, the hypertext author is free (and encouraged) to use high-level concepts and terminology. Novices can select the terms to learn about them while knowledgeable readers can move ahead to more complex topics. For example, in a historical database, key events, people, or places can be mentioned without description, and novices can follow the links to read the articles in related nodes if they need background material. A database on Austria and the Holocaust was based on people, places, events, organizations, and social organizations. These could be mentioned freely throughout the text and readers could follow the links to find more. Names of people or places that were not in the database were mentioned only when necessary and with a brief description.

Hypertext is conducive to the inclusion of appendices, glossaries, examples, background information, original sources, and bibliographic references. Interested readers can pursue the details while casual readers can ignore them.

Creating documents for a hypertext database introduces some additional considerations beyond the usual concerns of good writing. No list can be

---

* *Excerpt from Reflections on Authoring, Editing and Managing Hypertext, Chapter 8 from E. Barrett (Ed.) (1988)* The Society of Text, *Mass: MIT Press, © MIT Press. Reproduced with kind permission of the author and MIT Press.*

complete, but here again this list, derived from our experience, may be useful to others:

1. Know the users and their tasks: Users are a vital source of ideas and feedback; use them throughout the development process to test your designs. Realize that you are not a good judge of your own design because you know too much. Study the target population of users carefully to make certain you know how the system will really be used. Create demonstrations and prototypes early in the project; don't wait for the full technology to be ready.
2. Meaningful structure comes first: Build the project around the structuring and presentation of information, not around the technology. Develop a high concept for the body of information you are organizing. Avoid fuzzy thinking when creating the information structure.
3. Apply diverse skills: Make certain that the project team includes information specialists (trainers, psychologists), content specialists (users, marketers), and technologists (systems analysts, programmers), and that the team members can communicate.
4. Respect chunking: The information to be presented needs to be organized into small 'chunks' that deal with one topic, theme or idea. Chunks may be 100 words or 1000 words but when a chunk reaches 10,000 words the author should consider restructuring into multiple smaller chunks. Screens are still usually small and hard to read, so lengthy linear texts are not as pleasant. Each chunk represents a node or document in the database.
5. Show inter-relationships: Each document should contain links to other documents. The more links contained in the documents, the richer the connectivity of the hypertext. Too few links means that the medium of hypertext may be inappropriate, too many links can overwhelm and distract the reader. Author preferences range from those who like to put in a maximum of one or two links per screen, to the more common range of two to eight links per screen, to the extremes of dozens of selectable links per screen.
6. Be consistent in creating document names: It is important to keep a list of names given to documents as they are created; otherwise, it becomes difficult to identify links properly. Synonyms can be used, but misleading synonyms can be confusing.
7. Work from a master reference list: Create a master reference list as you go to ensure correct citations and prevent redundant or missing citations. Some hypertext systems automatically construct this list for you.
8. Ensure simplicity in traversal: Authors should design the link structure so that navigation is simple, intuitive, and consistent throughout the system. Movement through the system should be effortless and require a minimum of conscious thought. Find simple, comprehensible, and global structures that the readers can use as a cognitive map. Be sensitive to the possibility that the user will get 'lost in hyperspace' and develop the system so recovery is simple.

9. Design each screen carefully: Screens should be designed so they can be grasped easily. The focus of attention should be clear, headings should guide the reader, links should be useful guides that do not overwhelm the reader. Visual layout is very important in screen designs.
10. Require low cognitive load: Minimize the burden on the user's short-term memory. Do not require the user to remember things from one screen to another. The goal is to enable users to concentrate on their tasks and the contents while the computer vanishes.

Creating the introduction

A key design issue is how to organize the network and how to convey that order to the reader. Some documents begin with an Executive Overview that summarizes and provides pointers to sections. Some reference books have a main table of contents that points to tables of contents for each section or volume. Most books start with a hierarchical table of contents. These models can be a guide to authoring strategies for creating the root document.

1. Make the root document an overview that contains links to all major concepts in the database [glossary strategy].
2. Adopt a hierarchical approach in which the links in the root document are major categories [top-down strategy].
3. Organize the root article as a list or table of contents of the major concepts in the database [menu strategy].

The suitability of the different authoring strategies will depend upon the purposes and anticipated use of the database.

Article size: Small is beautiful (usually)

A major concern to authors of hypertext databases is determining the optimal length for documents. Research suggests that many short documents are preferable to a smaller number of long documents.

An experiment was performed at the University of Maryland using the Hyperties system in which the same database was created as 46 short articles (from 4 to 83 lines) and as 5 long articles (104–150 lines). Participants in the study were given 30 minutes to locate the answers to a series of questions by using the database. The 16 participants working with the short articles answered more questions correctly and took less time to answer the questions.

The optimal article length may be affected by such variables as: screen size, nature of task, session duration, and experience of user. One problem with databases consisting of many small articles is that it increases the amount of navigation the reader must perform.

Converting existing documents and files

Converting existing documents into hypertext form is a major concern of hypertext developers. Thousands of large online databases already exist and are available via information retrieval systems such as DIALOG, BRS, or

Nexis/Lexis. Putting these databases in hypertext format would be a monumental task. Links would need to be placed in each record (document) and browsing capabilities added to these databases. If links were to be established across databases, they would need to have comparable structures.

It seems likely that many existing databases will be converted to hypertext form (for example the Oxford English Dictionary and the AIRS Bible projects). In some cases, only new records added to databases will contain coded links suitable for hypertext.

In the personal computer domain, text conversion is much more feasible since most PC based hypertext systems accept standard ASCII files as input. Most existing documents can be converted to ASCII format. This leaves the task of identifying links using the authoring capabilities of the hypertext program.

Many documents to be converted contain various kinds of graphics. The conversion of graphics to hypertext format is problematic. Graphics file formats differ widely across systems. Modern digitizing technology makes it possible to convert most graphic images from paper to electronic form so they can be incorporated into hypertext databases. However, the degree of manipulation possible with the graphic once in electronic form (e.g., resizing, rotation, cropping, etc.) depends upon the graphic editor available.

There is good reason to hope that processes for automatic conversion will be widely developed. We have already succeeded in converting databases with explicit and consistent structure that contained in document formatting commands. The process involves writing a grammar and parser for the input and a generator to output the articles and the links.

Managing a hypertext project

Each project is different and each manager may have different styles of work, but again our experience may be useful for others. This list can surely be extended:

Identify application that satisfies the Golden Rules of Hypertext

- Large body of knowledge separable into smaller components
- Interrelated components
- User needs only a slice at any time

Design knowledge structure

- Specify goals, market niche & audience
- Decide on scope of coverage
- Identify list of topics and components
- Choose traversal structures

Prepare material

- Collect or create material
- Develop a style sheet for writing articles and creating links
- Ensure appropriate cross referencing to related concepts
- Arrange for editing of text and graphics

- Secure legal permissions
- Create database in proper formats
- Work with graphic artists to create images

Run test

- Insist on multiple reviews of the database
- Test hardware, software & database
- Test browsing and fact finding
- Capture usage data
- Revise and refine
- Prepare acknowledgements and credits

Dissemination

- Develop package design and installation instructions
- Start with small group and expand
- Provide consultation for problems
- Plan improvements

Each of the more than thirty hypertext databases we have built was different. We try to begin by forming a clear concept of the structure of the entire database and its sections. For example, in the database on Austria and the Holocaust, topics were chosen by lead historian Marsha Rozenblit who identified five categories: people, places, events, organization and social movements. We find it extremely helpful to write an initial list of proposed articles so that authors would have a good idea of which links might be added as they write their articles.

A key step is writing the introductory article which conveys the overall concept and points to the key articles in the database. These key articles point to each other and to secondary articles. The Table of Contents is a second chance to provide orientation for the reader and we generally revert to traditional indented formats found in most books. Adding a Table of Figures to the Table of Contents seems important since most readers of books and hyperbooks like to look at pictures. In many projects we use some articles as a sub-index to give a tabular list of articles on a common topic. For example, in the Guide to Opportunities in Volunteer Archaeology, there are lists of dig sites by geographic regions and by historical periods.

Once a few articles, say 10%, have been written it is important to try browsing the database to see if the organization is comprehensible and if the writing style is acceptable. Some projects began with long articles (*CACM* July 1988 Special Issue on Hypertext) which seemed more attractive when separated into smaller articles. Other projects (EDUCOM Conference Guide) consisted of almost entirely one-page articles. We regularly discuss the number of links per screen and take advice from reviewers and usability testers. A frequent policy is to highlight only on the first occurrence of a term in an article to reduce distracting clutter.

A devoted managing editor is necessary to move the writing along, coordinate with graphic artists, ensure that reviews are done, guarantee that copy editing and final fixes are performed diligently, and to handle the disk production in a timely fashion. Our projects were in the range of one to four month efforts by two to five people with additional consultations from reviewers. Sometimes projects stretched over more months if reviews took longer or if later changes were needed. The Hyperties author tool greatly facilitates productivity as we and several others have found: 'Overall Hyperties offered the best platform for creating links to references and pictures.' (Harris and Cady, 1988).

Acknowledgements and credits should be handled in an explicit and professional manner. The many participants, including the hypertext editors/authors, graphic artists, copy editors, reviewers, and programmers should be recognized. Hypertext, like movie production, can involve dozens of people and it is perfectly acceptable to have a long list with specific roles identified.

Final production details like the packaging design and manufacturing, contents of the disk labels, instructions in paper and disk formats, creation of installation programs, and coordination with distributors must all be handled carefully. An excellent effort on the contents becomes lost if the packaging permits damage to the product or the user can't follow the instructions to load the database.

The psychology of hypertext

From the earliest literature on hypertext (e.g., the July 1945 *Atlantic Monthly* article 'As We May Think' by Vannevar Bush), much emphasis has been placed on the idea that hypertext structures data in a manner similar to human cognition: in particular, the organization of memory as a semantic network in which concepts are linked together by associations.

If this is valid, it suggests that hypertext should be an efficient way of learning. Learning theory would predict that hypertext should improve meaningful learning because it focuses attention on the relationships between ideas rather than isolated facts. The associations provided by links in a hypertext database should facilitate remembering, concept formation, and understanding.

In addition, the greater sense of control over the reading process may produce increased involvement and desire to read more. In the same way that computer games can be very absorbing because of the high level of interactivity, hypertext databases may be very engaging too.

Getting started

Once you have a feel for hypertext, the next step is to experiment with creating your own hypertext documents. You will need to obtain an authoring system and might start with something familiar such as your Hyper-resume, especially if you already have the contents in machine-readable form. Another modest start would be a personal autobiography or a family newsletter or family tree in hypertext form.

More ambitious projects (a day or two of work) might be to implement part of a personnel policy database where you work, a community Hyper travel guide for your neighborhood (restaurants, stores, emergency services, etc.), or maybe a personal database of your cassette tapes, antiques, or books. These projects would compel you to organize the knowledge in some structured form, recognize relationships within and across groups of nodes, identify the central ideas that would become links from the root document, and decide how to use graphics.

Once you are satisfied with the modest project you can move on to a major project that might occupy you for several weeks or months. There are many attractive candidates and some of them could become viable commercial ventures. Repair manuals, training manuals, advertisements, corporate annual reports, organization trees for large companies, travel guides, sports and entertainment databases, and self-help guides are all possible. And just for fun why not mystery novels, joke books, and adventure games. Let your imagination be your guide!

Acknowledgements

Hyperties has been under development since 1983 and many people have participated in its design and refinement. Dan Ostroff created the initial versions in APL, then converted to C, and has remained an important influence. Janis Morariu and Charles Kreitzberg provided valuable design guidance. Paul Hoffman and James Terry of Cognetics Corporation have worked hard to convert our software to a solid commercial product. Many others participated in developing various versions on the IBM PC and the SUN including major contributions from: Kobi Lifshitz, Richard Potter, Bill Weiland, Don Hopkins, Rodrigo Botafogo, and Catherine Plaisant-Schwenn.

Initial funding came from the US Department of Interior in relation to the US Holocaust Memorial Museum and Education Center, where David Altshuler and Anna Cohn were early supporters of our vision. Later support for our user interface and hypertext work came from Apple, AT&T, IBM, Museum of Jewish Heritage, NASA, and NCR. We are grateful to all the institutions and individuals that recognized the importance of our efforts.

Parts of this chapter were drawn from Hypertext Hands-On!, written with Greg Kearsley and published by Addison-Wesley, Copyright 1989.

References

Harris, Margaret and Cady, Michael (1988, November). The dynamic process of creating hypertext literature, *Educational Technology*, 27, 11, 33–40.

Koved, Larry (1985, July). Restructuring textual information for online retrieval, Unpublished Masters Thesis, Department of Computer Science, University of Maryland Technical Report 1529 (CAR-TR-133).

Kreitzberg, Charles and Shneiderman, Ben (1988). Restructuring knowledge for an electronic encyclopedia, *Proc. 10th Congress of the International Ergonomics Association*.

Weldon, L. J., Mills, C. B., Koved, L., and Shneiderman, B. (1985). The structure of information in online and paper technical manuals, *Proc. Human Factors Society—29th Annual Conference*, Santa Monica, CA, 1110–1113.

(Suggested further reading is given at the end of the next chapter)

Chapter 15

Restructuring Knowledge for an Electronic Encyclopedia*

Charles B. Kreitzberg and Ben Shneiderman

Introduction to hyperties^{tm}

Hyperties is a powerful, yet simple, new software tool for organizing and presenting information. It has been developed over the past five years at the University of Maryland's Human-Computer Interaction Laboratory and has been used for more than 50 projects (Shneiderman 1987a, 1987b). Hyperties authors can create databases consisting of *articles* that contain text and illustrations. Without the need for programming, authors can link these articles together so readers can easily browse through them.

Hyperties can be used for a wide variety of applications, including:

- On-line encyclopedias
- Newsletters
- On-line help
- Instruction and dynamic glossaries
- Reference manuals
- Corporate policy manuals
- Summaries of products and services
- Employee orientation
- Biographies
- Regulations and procedures
- Museum exhibits

The strategies for gaining the benefits of paper texts are well understood, but there is a great need for study of how knowledge must be restructured to take advantage of hypertext environments (Yankelovich, Meyrowitz & Van Dam, 1985; Conklin, 1987; Marchionini & Shneiderman, 1988). This chapter provides some guidance for designing Hyperties databases and reports on an exploratory study of comprehension tasks when article length was varied.

---

* *Paper from* Ergonomics International 88 - Proceedings of the 10th Congress of the International Ergonomics Association, *Taylor & Francis, London, 1988, © Ergonomics Association of Australia, Inc. Reprinted with kind permission of the authors and the Ergonomics Society of Australia, Inc.*

Authoring and browsing

Hyperties consists of two programs: *The Authoring System* and *The Browser*. The Authoring System is used to create a database of articles and illustrations. Using the authoring system is simple—like using a familiar word processor. The author types in the text of the articles and specifies the links or cross references

```
WASHINGTON, DC: THE NATION'S CAPITAL                    PAGE 2 OF 3

     Located between Maryland and Virginia, Washington, DC
   embraces the White House and the Capitol, a host of
   government offices as well as the Smithsonian museums.
   Designed by Pierre L'Enfant, Washington, DC is a graceful
   city of broad boulevards, national monuments, the rustic
   Rock Creek Park, and the National Zoo.

     First-time visitors should begin at the mall by walking
   from the Capitol towards the Smithsonian museums and on
----------------------------------------------------------------
SMITHSONIAN MUSEUMS: In addition to the familiar castle and
popular Air & Space Museum there are 14 other major sites.
FULL ARTICLE ON "SMITHSONIAN MUSEUMS"

NEXT PAGE  BACK PAGE   RETURN TO "NEW YORK CITY"     INDEX
```

Figure 15.1. This Hyperties display on an IBM Personal Computer shows highlighted embedded menu items that can be selected by touch screen or arrow keys. The user can follow a topic of interest, turn pages (NEXT or BACK), RETURN to the previous article, or view the INDEX.

to other articles and illustrations. Hyperties automatically ties the articles and illustrations together into a unified database and constructs an index to the entire database (Figure 15.1).

The Browser enables readers to access the Hyperties database of articles and illustrations. Using the browser is extremely easy and requires virtually no training. Readers can access complete articles, definitions of important terms, illustrations and cross references by using only three keys: the left arrow key ←, the right arrow key →, and the enter key. If the computer is equipped with a touchscreen, readers can browse without the use of a keyboard at all.

Links: The power of Hyperties comes from the *links* that tie articles and illustrations together. A link is a cross reference, an indication that more information on a particular word or phrase is available. For example, suppose you were writing an article on the joys of owning pet fish. In a Hyperties article, you might write a sentence such as:

> Among the most interesting fish are ~guppies~ and ~goldfish~.

The tildes (~) that surround the words *guppies* and *goldfish* inform Hyperties that these words are links to additional information. The additional information may be simply a definition or footnote. Or it may be a complete article, with links of its own. When the Hyperties browser displays text, the tildes are removed and the links are highlighted on the computer screen:

> Among the most interesting fish are **guppies** and **goldfish**.

On a monochrome monitor (IBM PC or compatible, or SUN 3 workstation, the links are displayed in boldface, like the example above. On a color monitor the links are displayed in a different color from the rest of the text. The highlighted text signals the reader that more information is available.

The reader may choose to explore the database by using the links to travel among articles and illustrations. Hyperties automatically keeps track of the path so readers can return to previously seen articles.

Illustrations: Illustrations for Hyperties databases are prepared using a standard graphics editor. A scanner can be used to capture photographs and drawings. Prototype versions of Hyperties also support videodisk.

Index: Sometimes readers will be looking for a specific article and will not want to browse the database starting with the introductory article. Hyperties automatically creates an index which lists all the articles in the database. Readers may go to the index at any time and access any article in the database directly.

Synonym: Authors may wish to refer to the same article using different words or phrases as links. This is often a matter of style. For example, suppose you were creating a Hyperties database about the presidents of the United States and included an article on George Washington. Here are three sentences you might write (in the same or different articles) that link to the same article on George Washington:

> **George Washington** was the first president of the United States.
>
> A well-known anecdote about **Washington** involves a hatchet, a cherry tree, and his father's wrath.
>
> In understanding the political motivation of **the president**, it is important to consider his roots.

Hyperties can treat **George Washington, Washington**, and **the president** as synonyms which all link to the same article. The author need not plan this in advance; as Hyperties builds its index it will ask if certain terms are to be considered synonyms or not.

Highlighting is Selective: Just because a word or phrase can be used as a link to an article or illustration, does not mean that it must be used that way every time it appears. Authors decide when a word or phrase is a link by enclosing it in tildes. This keeps article from becoming cluttered with gratuitous highlighted terms. For example, you could write:

> ~George Washington~ was a military leader and as such commanded political respect. It was said that George Washington was autocratic and it was said that George Washington was a democrat.

When Hyperties displays this text, only the first reference would be highlighted as a link:

> **George Washington** was a military leader and as such commanded political respect. It was said that George Washington was autocratic and it was said that George Washington was a democrat.

Introductory Article: In each Hyperties database the author specifies an article as the lead or introductory article. Since many encyclopedia readers will browse the introductory article first, this article should be composed so that it references as many key articles as possible. There are several strategies for composing the introductory article.

One strategy is to fill the introductory article with many references making it a summary of the entire database. By scanning it, the reader can select one of many places to begin browsing.

A second strategy for the introductory article is to confine it to only a few key references. In this strategy, the idea would be to minimize the number of details which the reader must deal with and start him or her down an appropriate path. For example, suppose you were building a policy manual which had many detailed articles on specific policies. Rather than referencing many policies in the introductory article you could develop a more general approach such as the following:

> This database contains policies relating to:
>
> **permanent employees, temporary staff**, and **consultants**. In addition you will find policies which apply to all staff relating to **security, non-disclosure**, and **dealing with the press**.

A third strategy for the introductory article is to design it as a high level index. Here is an example:

CORPORATE POLICIES

Permanent Staff
- **Hiring Permanent Staff**
- **Termination Procedures**
- **Benefits and Vacation Policies**

Temporary Staff
- **Approval Policy for Hiring Temps**
- **Approval Policy for Retaining Consultants**

This technique of using the introductory article as an index can be extended to other articles. For example, **Approval Policy for Hiring Temps** could link to a new article that provided a detailed 'index' to relevant policies, for example:

APPROVAL POLICIES FOR HIRING TEMPS

Hiring Short-term Temporaries from Agencies
Establishing Qualifications of Temporary Staff
Hiring Independent Contractors

This technique can yield an extensive network of indexes. A particular article could appear in several indexes, so readers can access it from many points. For example, an article on *Vacation Policies for Permanent Staff* could be highlighted under *vacations, benefits, permanent staff*, or any other relevant area.

Planning for Expansion: The Hyperties browser will not highlight a reference to an article or picture unless the article or picture exists. If you are writing an article and are discussing a topic which may eventually be the subject of its own article, you can put tildes around a link to a word or phrase. Because the references article does not exist, Hyperties will not highlight the link. Later, when the article is written, Hyperties will automatically highlight the reference.

Writing style

In general, it is best to keep articles short, and keep a sharp focus. Instead of discussing a subsidiary topic which is not the main subject of an article, you can merely mention it, delimiting the key word or phrase with tildes. Then you can make that topic the subject of its own article, or at least give the topic a definition which can be called up by the reader.

The same technique can be applied to details. Rather than including detailed information in an article you can simply reference it and create separate articles for it. This shields the reader from unnecessary details, but provides a path to them when the reader deems it relevant. This technique can be especially useful when the material contains case studies, experiments or many examples.

Creating instructional material

Hyperties can be used as a tool to reduce the difficulty of creating educational software and allow authors to focus on content and instructional design, rather than on technical factors. In Hyperties, concepts and information can be entered and linked together. Developing courseware in such an environment is more like writing a book than writing a computer program. With Hyperties, the development of courseware should become an instructional, rather than technical, endeavor.

For example, an introductory psychology module might contain the following text:

> The basic process in **behavioral psychology** was presumed to be **conditioning**. Two types of conditioning were extensively studied: **operant conditioning**, the more powerful form is most associated with **B. F. Skinner**; the less powerful paradigm of **respondent conditioning** is most frequently associated with the studies of **Ivan Pavlov**.

The highlighted words and phrases in boldface indicate to the student that additional information is available on the topics: **behavioral psychology, conditioning, operant conditioning, B.F. Skinner, respondent conditioning**, and **Ivan Pavlov**. This additional information might be: a definition (for example, a definition of behavioral psychology), a new article (for example, an article on operant conditioning), an illustration (for example, a graph of conditioning and extinction), or a videodisk sequence (a brief biography of Ivan Pavlov).

Implicit within Hyperties is a cognitive model based on associative relationships. Articles in Hyperties explain concepts and tie articles, illustrations, and videodisk sequences together to create relationships among concepts. A Hyperties database may therefore be viewed as an associative network of concepts and examples at various levels. The power of this simple structure is attractive. Instructors can express the relationships among ideas by the manner in which articles are linked. Concepts can be expressed at multiple levels, with high level concepts linked to more specific concepts and specific concepts linked to examples.

Because a Hyperties database is organized according to the relationships inherent in the instructional material, it may help students learn the material in an integrated, holistic fashion. One of the most difficult instructional tasks in any content area is conveying the systems and interrelationships which underlie the facts. Memorization of isolated facts leads to rote learning; integration of concepts and their relationship into the learner's cognitive structure should lead to meaningful, useful learning.

Hyperties also creates materials that are learner-controlled. Much computer-based instruction is based on a dialogue model in which the computer constantly prompts the student to respond to questions. This can be a powerful instructional model but is often implemented in a fashion which forces the student to accommodate to the pace and presentation units of the software. It is this factor which makes so much computer-based education unappealing to the student. In Hyperties, the student has greater control. Learning proceeds according to the pace and paths selected by the student. Students need not waste time on material they already know and they can pursue a topic of interest to any depth desired.

We do not suggest that Hyperties is ideal for all forms of instruction; rather, we suggest that it is excellent for certain types of instruction. In particular, Hyperties, we believe, will be extremely effective in the following areas:

- *familiarization*—situations in which a person is introduced to a new content area and needs to become familiar with the key concepts which underlie it and their relationships.
- *annotation*—situations in such areas as literature, poetry, art, law, and politics in which students read source material which may be heavily annotated. The annotations remain 'hidden' behind the links and so do not interrupt the flow of primary material but are instantly available for reference.
- *dynamic glossaries*—glossaries prepared in Hyperties not only define terms but provide links to related concepts. This enables the reader to more fully understand the key terms.
- *diagnosis and review*—coupled with objective test items, Hyperties becomes a powerful diagnostic and review technique for any achievement test. Students answer questions. If they are correct, the software moves on to the next item. If the student answers incorrectly, Hyperties presents an explanation which serves as an entry to the database. Students can browse the database until they feel confident they understand the area and then return to the diagnostic test.

- *diagnostic problem-solving*—maintenance problems fit the Hyperties structure conveniently. The reader can select model numbers, problem features, symptoms, or other conditions and receive further information about how to proceed. This strategy has potential for machine repair, business procedures, medical diagnosis, etc.
- *organizational information*—when there is a need to teach organizational relationships (for example, the structure of government) Hyperties can represent the relationships by linking informational articles together. Hyperties is also useful in presenting the facilities and services of an organization. For example, it can be used to create orientation courses for new college students.

Newsletters

Hyperties can be used to create efficient newsletters. The essence of a newsletter is that the reader wants to obtain up-to-date information efficiently. In Hyperties, you can create a series of 'headlines' or short abstracts which let the reader know what information is available. If the information is of interest, the reader can then select the article for more information. For example, the first page of a newsletter for personnel departments might be structured as follows:

> NEW COURT RULING ON MATERNITY LEAVE. A federal court recently ruled that corporations must provide maternity leave to long term temporary employees. See **Higgens vs Retco**.
>
> LIABILITY ON ALCOHOL-RELATED ACCIDENTS. An employee, who became drunk at a company sponsored party and later was injured in an auto accident, sued the company. For details of this case see **Carnevale vs Rapido Trucking**. For a review of the legal issues see **Alcohol and Corporate Liability**.

Creative reading sequences

A final authoring strategy is the development of reading sequences in Hyperties. By its nature, the articles in a Hyperties database can be randomly accessed. However, sometimes it is desirable to provide readers with a path to read the articles in sequential order.

For example, suppose a set of materials were organized into five key articles. These articles contain the main ideas that your readers should encounter. In addition, there are a number of more detailed articles, that expand upon the key articles. What you want the reader to do is:

(1) read all five key articles in sequence

(2) within a key article, use the browser to explore the details of any concepts for which (s)he wants more information.

You can accomplish this in at least two ways. The first way would be to use the introductory article to list the five key articles:

Please read the following key articles, in sequence:

Article 1: **Buying your Fishtank**

Article 2: **Selecting your Fish**

Article 3: **Setting Up Your Tank**

Article 4: **Feeding Your Fish**

Article 5: **Care of Fish Babies**

The reader would select each article in turn. Within each article, the reader could browse related articles of interest and, ultimately, return to the introductory article to select the next article.

The second technique is to create a link at the end of each article to the next article which you want the reader to see. For example, the first article, Buying Your Fishtank, might end as follows:

> Congratulations! Having followed the instructions in this article, and you are now the proud owner of a fishtank. But what is a fishtank without fish? To move on to this exciting step, please read **Selecting Your Fish**, next.

This technique can be expanded to create several paths through the material, if you desire.

Article length: an exploratory experiment

One design issue about this environment is the appropriate size of articles. Since following a reference by turning to another article is more rapid in Hyperties than on paper, smaller fragments of text may be more suitable. Also, considering the small size of many computer screens, slow page turning, and possibly poor readability, some designers suggest that hypertext articles should be brief. To test this conjecture an exploratory study was run by Dana Miller and Anna Williams under the direction of the second author.

Thirty-two psychology student volunteers were given brief instructions and ten minutes of practice with a Hyperties database dealing with personality types. Then the subjects were given 30 minutes to answer as many multiple-choice questions as they could. A typical question was: What are introverts interested in at their work? (A) the procedures. (B) the results. (C) the idea behind the job.

The major independent variable was article size. The short database had 46 articles from 4 to 83 lines, while the long database had five articles of 104 to 150 lines. A second independent variable was personality type of the subjects, but the measurement instrument was ad hoc and this variable failed to produce statistically significant results. Therefore the remainder of the discussion focuses on the article length variable and performance time data.

The 16 subjects working with the short articles answered an average of 10.1 questions correctly while the 16 subjects working with the long articles answered an average of only 7.2 questions correctly ($F(3,30)=4.73$, $p<.01$). *The average time per correct answer was 125 seconds with the short articles and rose to 178 seconds for the long articles* ($F(3,30)=9.22$, $p < .001$).

While these results support the conjecture that short articles facilitate fact finding in a hypertext environment, they need replication with other databases, questions, subjects, and screen environments (e.g. larger or multiple windows). Longer practice sessions, an effective subjective satisfaction questionnaire, and within subjects design would be useful improvements in a new study.

Conclusion

The new opportunities offered by hypertext systems call for a re-thinking of the presentation of knowledge. Hyper-chaos is a serious danger and effective guidelines will be necessary to assist authors in creating useful materials. Empirical studies can be a great aid in sharpening thinking, developing appropriate guidelines, and formulating new theories of how people seek and acquire knowledge.

References

Conklin, Jeff (1987, September). Hypertext: An introduction and survey, *IEEE Computer* 20, 9, 17–41.

Marchionini, Gary and Shneiderman, Ben (1988, January). Finding facts vs. browsing knowledge in hypertext systems, *IEEE Computer* 21, 1, 70–80.

Shneiderman, Ben (1987a). User interface design for the Hyperties electronic encyclopedia, *Proc. Hypertext '87 Workshop*, 199–204, Raleigh, NC: University of North Carolina.

Shneiderman, Ben (1987b). User interface design and evaluation for an electronic encyclopedia, In: Salvendy, G. (Ed.) *Cognitive Engineering in the Design of Human-Computer Interaction and Expert Systems*, 207–233. Amsterdam: Elsevier.

Yankelovich, N., Meyrowitz, N., and Van Dam, A. (1985, October). Reading and writing the electronic book, *IEEE Computer*, 19, 10, 15–30.

Suggested further reading

Barrett, E. (Ed.) (1988). *Text, Context and Hypertext*. Cambridge, MA: M.I.T. Press.

Barrett, E. (Ed.) (1989). *The Society of Text*. Cambridge, MA: M.I.T. Press.

Begoray, J. (1990). An introduction to hypermedia issues, systems and application areas. *International Journal of Man-Machine Studies*, 33, 2, 121–148.

Communications of the ACM Vol. 31, No. 7, July 1988. (Special issue on hypertext).

Conklin, E. J. (1987). Hypertext: an introduction and a survey. *IEEE Computer*, 21, 17–41.

Hartley, J. (1990). Hype and hypertext. *Higher Education*, 20, 113–119.

Horn, R. E. (1989). *Mapping Hypertext*. Information Mapping, 303 Wyman Street, Waltham, MA 02154, USA.

Jonassen, D. H. (1989). *Hypertext/Hypermedia*. Englewood Cliffs, NJ: Educational Technology Publications.

McAleese, R. (Ed.) (1989). *Hypertext: Theory and Practice*. Oxford: Intellect Books.

McAleese, R., & Green, C. (Eds.) (1990). *Hypertext: State of the Art*. Oxford: Intellect Books.

Nielsen, J. (1989). Hypertext bibliography. *Hypermedia*, 1, 1, 74–91.

Rada, R., Keith, B., Bourgoigne, M. & Reid, D. (1989). Collaborative writing of text and hypertext. *Hypermedia*, 1, 2, 93–110.

Raymond, D. R. & Tompa, F. W. (1988). Hypertext and the Oxford English Dictionary. *Communications of the A.C.M.* 31, 7, 871–879.

Shneiderman, B. & Kearsley, G. (1989). *Hypertext—Hands On!*, Reading, MA: Addison Wesley.

Shneiderman, B., Brethauer, D., Plaisant, C. & Potter, R. (1989). Evaluating three museum installations of a hypertext system. *Journal of the American Society for Information Science*, 40, 3, 172–182.

Wright, P. (1987). Reading and writing for electronic journals. In Britton, B. K. & Glynn, S. M. (Eds.) *Executive Control Processes in Reading*, Hillsdale, NJ: Erlbaum.

Wright, P. (1989). Interface alternatives for hypertexts. *Hypermedia*, 1, 2, 146–166.

Part V

Computer-aided writing: future trends

Chapter 16

Designing Idea Processors for Document Composition*

Ronald T. Kellogg

In this chapter, I discuss major problems that people experience during the prewriting and drafting stages of document preparation, and describe computer tools, generically called idea processors, that may alleviate these problems. Such work diverges from the mainstream of psychological research on computers. Previous research on human-computer interaction has focused on reducing the complexity of the interface to enhance the usability of the machine. Although improved interface design is a valuable goal for human-computer research, it should not be the only goal. Landauer (1985) recently suggested that psychology might also be fruitfully applied to the problem of conceiving new cognitive tools that might be useful to people. Understanding the limitations facing a person in a specific task, such as composing, and then formulating the type of tools that might aid performance are also important areas for psychological research.

My goals here are (1) to indicate why idea processors are important cognitive tools to be explored as composition aids, (2) to detail three fundamental limitations encountered by even experienced writers during the prewriting and drafting of text, (3) to formulate three relatively novel computer functions that could assist writers, illustrating these with relevant software, and (4) to consider the potential effectiveness of such functions in overcoming the weaknesses of writers.

Why idea processors

Idea processors are programs that assist the writer in generating and organizing ideas or concepts so that they can be communicated successfully to others. Such programs may be useful in several tasks, but the present analysis is limited to the task of planning ideas for a document and translating those ideas into text. Most of the idea processors under consideration are specifically tailored for writing, and writing has been the object of considerable psychological research in recent years.

The processes of prewriting, doing a first draft, and revising subsequent drafts include numerous activities and can require long periods of time to complete. It is useful to categorize these activities as collecting information

---

* *Reprinted from* Behavior Research Methods, Instruments and Computers *1986, 2, 118-128, © Psychonomic Society, Inc. Reprinted by kind permission of the author and the Psychonomic Society Inc.*

(reading, listening, and searching bibliographic sources), planning text (creating ideas, organizing ideas, and setting goals), translating plans into text (constructing legitimate sentences—actual language production), and reviewing text (reading, evaluating, editing errors). A well-documented fact is that collecting, planning, translating, and reviewing generally do not occur in a simple linear sequence. Instead, the processes occur recursively during prewriting and on drafts. Detailed theoretical accounts of these processes and evidence on their recursive nature are available from several sources (Beaugrande, 1984; Flower & Hayes, 1980b; Gould, 1980; Nold, 1981).

Collecting, planning, translating, and reviewing are all amenable to computer assistance (Kellogg, 1985). DIALOG (Seymour, 1984), SCI-MATE (Garfield, 1983), and other bibliographic search systems are widely used tools for collecting information. BOSS (Walker, 1984) and numerous other spelling checkers are growing in popularity as reviewing aids. Writer's Workbench is a versatile tool for reviewing diction, punctuation, readability, and style, as well as spelling (Macdonald, 1983). In contrast to the relatively abundant selection and common use of tools for collecting and reviewing, tools for planning and translating are few in number and confined in use. Yet tools for planning and translating are clearly needed.

A series of experiments on letter writing (Gould, 1980) indicated that executives spend about two thirds of their writing time involved in planning ideas, most of the remaining time translating ideas into text, and relatively little time reviewing what they have written. Other experiments, using different measurement techniques and subjects, indicated that planning and translating combined consume about 75% of writing time (Kellogg, 1984). Getting started on a draft was reported to be the most difficult part of writing by 30% of the academic writers surveyed in one study (Green & Wason, 1982). In the extreme case of a blocked writer, planning and translating are so onerous that few if any words are ever produced for the composer to review (Boice, 1983).

Designing idea processors that effectively aid in the processes of planning and translating requires a theoretical analysis of the writer. Why do adult writers, who possess both the verbal skills and the motivation needed to write acceptable prose, need to invest so much time and effort in planning and translating? The answer may lie in three fundamental limitations of human thinking ability.

Fundamental limitations

Attentional overload

The first limitation is that writers experience difficulties in planning and translating when they simultaneously attend to reviewing. Limited attentional capacity is overloaded when writers attempt to juggle too many operations. Attention may be divided simultaneously among processes, rapidly alternated among them, or primarily focused on one process while others are executed automatically. Difficulties arise when insufficient time and effort are devoted to planning and translating because of competition from other processes. Effective planning and translation presumably require sustained concentration. Consequently, overloading attention by trying to review at the same time leads to poor planning and translating. This limitation has been discussed in theoretical accounts of composition (Elbow, 1981; Green & Wason, 1982), and three sources of evidence can be marshalled to support it.

First, verbal protocols of college students thinking aloud while composing reveal the problem of attentional overload (Flower & Hayes, 1980b). In commenting on their protocol analyses, Flower and Hayes noted that

> Writing is the act of dealing with an excessive number of simultaneous demands or constraints. Viewed this way, a writer in the act is a thinker on a full-time cognitive overload . . . A writer caught in the act looks much more like a very busy switchboard operator trying to juggle a number of demands on her attention and constraints on what she can do (p. 33).

Second, an experiment by Glynn, Britton, Muth, and Dogan (1982, Experiment 1) showed that the quality of planning is adversely affected when writers attempt to translate and review at the same time. They examined the number of arguments generated by students in a persuasive writing task while manipulating via the instructions the number of processes juggled. The unordered-propositions condition encouraged the writers to focus their attention exclusively on generating ideas. The ordered-propositions condition prompted them to generate and organize their ideas. The mechanics-free condition added the requirement of translating their organized ideas into rough-draft sentences. Lastly the polished-sentences conditions encouraged the students to carefully review their sentences while they were translating and planning. An analysis of the total number of arguments produced on a preliminary draft revealed a significant effect of instructional condition. The means in the four conditions were as follows: polished sentences, 3.3; mechanics-free sentences, 4.9; ordered propositions, 8.3; unordered propositions, 13.3.

Third, direct measurements of the degree of effort given to planning, translating, and reviewing reveal the demanding nature of these processes. In one experiment (Kellogg, 1986, Experiment 1), I employed directed introspection to track when the writer attended to planning, translating, and reviewing, and secondary-task reaction times to measure the degree of effort given to each process. Collecting was not examined because the subjects were required to write from memory only. In a single sitting, college students wrote a persuasive essay. On a variable-interval schedule, the subjects heard an auditory signal while writing. This was a signal for the subjects to say 'Stop,' as quickly as possible, and reaction times were recorded. The subjects were instructed to pay primary attention to their writing and to respond to the signal as rapidly as possible as a secondary task. After saying 'Stop,' the subjects pressed one of four buttons to indicate whether their thoughts at the moment of the signal reflected planning, translating, reviewing, or some other process unrelated to these. The subjects had been trained to identify their thoughts as belonging to one of these four categories.

The mean reaction times associated with planning, translating, and reviewing were 712, 663, and 705 msec, respectively. The mean of baseline reaction times, collected when the subjects were not writing, was 346 msec. An analysis of variance revealed that baseline reaction times were significantly less than the times associated with the three writing processes [$F(3,84) = 92.56$, $p < .001$].

An interference difference score (writing process minus baseline) is an index of how much effort was given to each process: the greater the interference, the more demanding the task. The data for the writing task are plotted on the left side of Figure 16.1. Planning and reviewing showed the same large expenditure of effort; translating showed significantly less [$F(2,56) = 3.64$, $p < .05$]. To clarify the meaning of these interference scores, it is useful to compare them to those obtained in other experimental tasks. I draw on studies conducted in my laboratory and in other laboratories to make this clear. In all cases, the subject's primary task was complex, involving several cognitive processes, and a rapid, timed response to the primary task was not required. Hence, reaction times to the secondary task reflected the thinking demands of the primary task, rather than rapid response demands. College students served as subjects and an auditory secondary probe was employed in all cases as well. There were minor procedural differences in the studies (e.g., the intensity of the probe), but these factors presumably affected baseline as well as dual task reaction times. By looking at the interference difference scores, it is possible to make comparisons across the various studies (see Figure 16.1).

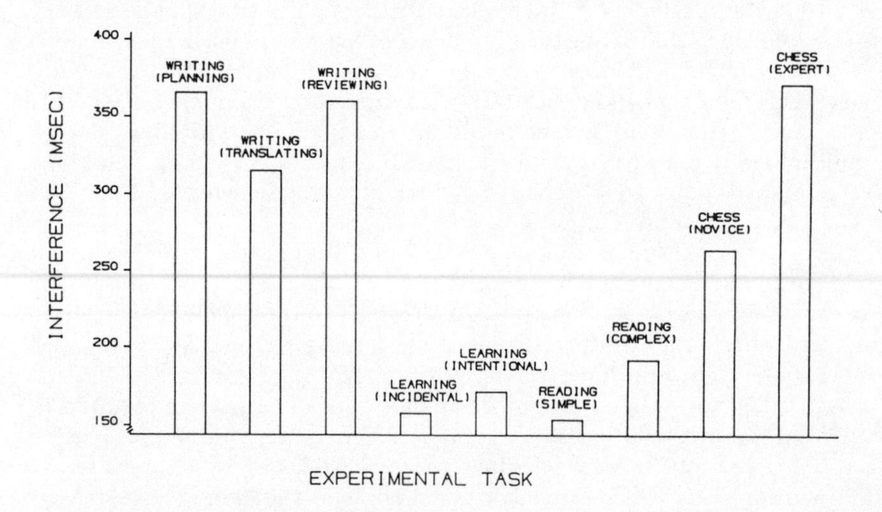

Figure 16.1 Secondary-task reaction time interference for various primary experimental tasks. Writing data are from Kellogg (1986), learning data are from Kellogg (1983), reading data are from Britton et al (1982), and chess data are from Britton & Tesser (1982).

In a study of my own (Kellogg, 1983) on learning a list of words, the interference scores for incidental learning and intentional learning instructions were less than half those obtained for the writing processes. Britton, Glynn, Meyer, and Penland (1982) reported relevant data on reading text of varying syntactic complexity; their interference scores both for reading simply syntax and for reading complex syntax were well below those observed for writing processes. Britton and Tesser (1982) examined the effort required to play chess. They had subjects determine the best move to make in several situations taken from the middle stages of an actual game. Their interference scores for novice players were less than, whereas those for expert players were slightly greater than those for writing. Thus, the effort demanded by planning, translating, and reviewing is substantial and is on the order of that required by expert chess play. It is easy to understand, therefore, how attentional overload can occur when the writer tries to juggle all three processes at once.

Idea bankruptcy

The second limitation often experienced by writers is a failure to generate usable ideas. Graesser, Hopkinson, Lewis, and Bruflodt (1984) noted that 'it is difficult for writers to generate ideas that are informative, interesting, sophisticated, and relevant to a particular pragmatic context' (p. 361). They aptly referred to this difficulty as 'idea bankruptcy'. This limitation of idea generation is perhaps self-evident to anyone who has tried to compose. It finds strong empirical support in the results reported by Graesser et al. (1984) and in the literature on creativity in problem solving.

Graesser et al. (1984) had college students write papers that exhibited their technical knowledge of economics, cancer, or growing flowers. The students wrote from memory for a reasonably long period of time (a minimum of 25 min), under instructions to write down everything that they knew on the topic. For example, the instructions in the economics condition were 'Write down all you know about the concept of economics, including inflation, recession, unemployment, and how they are related' (p. 346). Expert judges evaluated each statement on three dimensions (4-point scales). The truth dimension indexed the validity of the statement. The obscurity dimension measured whether the idea was a familiar piece of common knowledge. The sophistication dimension assessed how informative the statement was about relevant processes or mechanisms.

The mean number of statements was surprisingly low for economics (M = 9.6) and for growing flowers (M = 11.2), and was respectably high for cancer (M = 24.1). The overall mean truth score fared well (M = 3.06), indicating that most of the statements were true. However, sophistication and obscurity scores were uniformly poor across topics. The mean sophistication score was 2.45 on a 4-point scale, with a value of 1.00 indicating a very uninformative or irrelevant fact. Although most of the students had been exposed to relevant processes and mechanisms in course work, they rarely generated them in this task. The mean obscurity score was 1.61, with a value of 1.00 indicating a very popular or familiar idea (e.g., 'Cancer is a disease'). Graesser et al. (1984) summarized their findings as follows: 'Most of the generated ideas were true, culturally familiar (that is, not obscure), and unsophisticated' (p. 359).

Another source of evidence for the limitation of idea bankruptcy is the extensive literature on creative problem solving. Preparing a document can be fruitfully viewed as an ill-defined problem consisting of rhetorical, written-prose, and knowledge subproblems (Flower & Hayes, 1980a, 1980b). As in solving any problem, people are prone to errors in representing writing problems, searching the problem space for solutions, and evaluating tentative solutions (Hayes, 1981). The literature on creativity tests and on creativity training emphasizes the rarity of individuals who are fluent, flexible, and original in generating ideas (Guilford, 1967; Stein, 1974; Taylor & Barron, 1963).

Such classic impediments as functional fixedness (Duncker, 1945) and persistence of set (Luchins, 1942) undoubtedly contribute to idea bankruptcy in writing tasks. An inability to represent in novel ways objects, events, and concepts relevant to a writing problem might be viewed as a form of functional fixedness. Approaching every writing task in a routinized manner that has seemed to work well in the past illustrates persistence of set.

Affective interference

The third limitation experienced by writers is that fears and anxieties can interfere with successful composition. Powerful emotional reactions, both positive and negative, are commonly elicited by the process of writing (Green & Wason, 1982). These reactions are nicely reflected in a quotation attributed to the novelist James Jones: 'I hate writing. I love having written.' Lowenthal and Wason (1977) asked academic writers how they felt about the job, and most reported this type of mixed but intense response (e.g., 'Writing is a very hard grind—the good times come along only on the back of sweat and tears'). A few found nothing good about the experience (e.g., 'Writing is like being sick'), and a few took great pleasure in it (e.g., 'Writing is as enjoyable as making love'). Of concern here are the negative affective reactions that lead to procrastination, fretful attempts at writing, and complete avoidance of the task.

How serious is the problem of affective interference? For some writers, the affect spawned by writing may push them beyond the optimum level of arousal for so complex a task. Planning and translating may suffer when arousal levels are too high. If that is so, then writing quality and efficiency should suffer when writing anxieties are too high. Consistent with this expectation, Boice and Johnson (1984) observed a significant negative correlation between reported scholarly productivity and degree of writing anxiety among university faculty. Similarly, Daly (1978) investigated writing apprehension or anxiety among college students by developing an attitudinal questionnaire regarding anxiety and by correlating anxiety with writing performance. He concluded on the basis of several studies that writing-apprehensive students compose poorer quality documents than do less anxious students.

Writer's block occurs when fears and anxieties are so intense that would-be writers fail to begin or continue the process of composition. Rose (1980, 1984) found that blocked college students tended to follow rigid, maladaptive rules that disrupted successful writing. For instance, one blocker developed overly elaborate plans that lengthened the prewriting stage to several days. Then, with only a few hours left to create a first draft, the student found it impossible to translate the complex plan into a short essay.

Boice (1985) investigated differences in the self-talk of blockers and nonblockers among university faculty. His subjects recorded on note cards their thoughts during the initiation and completion of writing sessions. Over 5,000 examples of self-talk were collected and categorized. Boice identified seven categories of thoughts: work apprehension, procrastination, dysphoria, impatience, perfectionism, evaluation anxiety, and rules. Work apprehension (thoughts about the difficult, demanding nature of writing) and rules (thoughts about maladaptive formulas for writing, such as 'Good writing must be spontaneous and clever') occurred about equally often among blockers and nonblockers. Procrastination (thoughts that justify avoiding or delaying writing) was much more common among blockers (90%) than among nonblockers (55%). Dysphoria (thoughts reflecting burnout, panic, or obsessive worries), impatience (thoughts of achieving more in less time or imposing unrealistic deadlines), perfectionism (thoughts reflecting an internal critic who allows no errors), and evaluation anxiety (thoughts about fears of rejection) also afflicted blockers more than nonblockers.

More evidence of the seriousness of affective interference comes from Kubie's (1958) psychoanalytic treatment of creativity. Kubie challenged the popular belief that neurosis and creativity go hand in hand by describing cases in which creative productivity was diminished because of fear, guilt, and other anxiety states.

How common is emotional hindrance in writing? Procrastination seems to be universal, according to Green and Wason (1982). They found in their surveys of academic writers that getting started is judged to be difficult by all writers and is viewed as the single most difficult part of writing by 30%. Boice and Johnson (1984) found that 34% of their sample of university faculty reported moderate to high levels of anxiety about writing. A complete writing block, defined as an inability to write for some emotional/motivational reason, was reported by 12%. Rose (1984) reported that about 10% of the college population are blocked writers. Moreover, Freedman's (1983) survey of college students indicated that 45% found writing painful, 61% found it difficult, and 41% lacked confidence in their ability to write. Affective interference, therefore, seems to be a relatively frequent, as well as serious, difficulty.

Computer functions

In this section I describe three relatively novel computer functions that address the fundamental limitations detailed above. Existing programs are categorized as examples of programs that fulfil the function of funnel, inventor, or therapist. My purpose is to review the approaches and methods already taken to achieve these functions, with the hope of stimulating new design research in the area of idea processors for document composition.

Funnel

To help writers with attentional overload, a computer should serve the function of a funnel. I define a funnel device as an idea processor that channels the writer's attention into only one or two processes. By encouraging the writer to temporarily ignore reviewing and possibly translating, the funnel device might relieve attentional overload. There are several ways to design a funnel device. Currently available funnel devices adopt one of two approaches: distracting information or encouraging free writing (see Table 16.1). The first approach is seen in programs that allow the writer to expand and collapse an outline and in programs that selectively display the topic sentences of a document.

Outlining software performs the function of a funnel by allowing the writer to construct and retrieve a document at different levels of a hierarchical structure. For instance, to plan the main ideas of a document, without concern for translating or reviewing those ideas, the writer could collapse the outline and

Table 16.1. Idea Processors Serving as Funnel, Inventor, or Therapist

Approach	*Method*	*Software*	*Reference*
		Funnel Programs	
Hiding Distractions	Expand & Collapse Outline	NLS	Uhlig et al. (1979)
		Thinktank	Hershey (1984)
		Framework	Layman (1984)
		Promptdoc	Owens (1984)
	Topic Sentences Only	Writer's Workbench	Macdonald (1983)
		WANDAH	Von Blum & Cohen (1984)
Free Writing	Paced Writing	WANDAH	Von Blum & Cohen (1984)
		WRITER'S HELPER	Wresch (1984)
	Invisible Writing	WANDAH	Von Blum & Cohen (1984)
		Inventor Programs	
Creating Concepts	Topics, Pentad, Tagmemics	INVENT	Burns (1984)
	Visual Synectics	(Unnamed)	Rodrigues & Rodrigues (1984)
	Problem Statements	DRAFT	Neuwirth (1984)
	Nutshelling	WANDAH	Von Blum & Cohen (1984)
	Morphological Analysis	BRAINSTORMER	Bonner (1984)
Relating Concepts	Networks	(Unnamed)	Smith (1982)
	Trees	WRITER'S HELPER	Wresch (1984)
		Therapist Programs	
Covert, Embedded Therapy	Positive Reinforcement	INVENT	Burns (1984)
	Suggestion	INVENT	Burns (1984)
Overt, Independent Therapy	Cognitive Behavior	MORTON	Selmi et al. (1982)
	Contingency Management	(Unnamed)	Boice (1982)

display only the superordinate levels of the outline, hiding all subordinate points. The subordinate points might easily distract the writer from giving full attention to the superordinate levels. Thus, outlining programs explicitly encourage the writer to concentrate only on high-level planning in this example. Alternatively, to focus on translating a specific subordinate idea, the writer could hide all superordinate levels and expand only the subordinate point of interest at the moment. Once a subordinate point is completely translated, it can be selectively displayed for reviewing as well.

Outlining programs do not force the writer to plan first, translate second, and review third, in a linear sequence. On the contrary, they are highly compatible with a recursive strategy of mixing processes in various orders as needed. The writer can certainly shift from, say, reviewing a paragraph stored as a subordinate point to planning a new idea at the highest level of the hierarchy. However, by hiding distracting text, such programs do help the writer to finish reviewing the subordinate point before advancing to planning the new superordinate point.

NLS is an example of a program that expands and collapses outlines (Uhlig, Farber, & Bair, 1979). It uses an infinitely deep outline structure to organize an evolving text. Each level consists of text ranging in length from a single word to an entire paragraph. The writer develops the outline by adding levels to a hierarchical structure in any manner desired. For instance, the writer might start with three superordinate ideas that are labeled by NLS as 1, 2, and 3. Next, the writer thinks of a subordinate 4. Finally, he thinks of a subordinate to idea 1A, and NLS labels it 1A1. The writer can view the outline in different ways, by selectively hiding information. By collapsing and expanding the outline, the writer may more easily focus attention on one or two processes at a time. Thinktank (Hershey, 1948), Framework (Layman, 1984), and Promptdoc (Owens, 1984) are similar to NLS. Each differs from NLS in the details of how the system numbers the text entries and how the outline is expanded and collapsed.

A second method of hiding distractions is to display only the topic sentence of each paragraph of a document. This method is useful for planning or reviewing the macro-structure of a text while ignoring the details. Writer's Workbench (Macdonald, 1983) displays the first and last sentence of each paragraph. WANDAH (Von Blum & Cohen, 1984) selects the first sentence or any sentence specifically designated by the writer as a topic sentence.

Free writing refers to rapid translation, following whatever meager plan is available without concern for extensive planning or reviewing (Elbow, 1981). It involves quickly writing off the top of one's head in a free-association, stream-of-consciousness manner. The aim of free writing is to put one's thoughts on paper before one's internal editor rejects them as unsophisticated or lacking in style. The product of free writing can and must be scrutinized and edited at a later time. A closely related technique is *brainstorming* (Flower & Hayes, 1977), which is more goal directed and less free associational than free writing. In brainstorming, the writer either starts with a plan or develops one while composing, but reviewing is forbidden, as in free writing. Here I use the term free writing to cover both Elbow's (1981) usage of the phrase and the technique of brainstorming.

WANDAH (Von Blum & Cohen, 1984), a software package designed for university-level writing classes, includes programs that promote free writing. WANDAH encourages free writing by flashing the screen when the writer pauses for too long (long pauses indicate that the writer is planning or reviewing). The flashing serves as a funnel by reminding the writer to focus on translating rapidly. A similar approach is taken in Wresch's (1984) free-writing program that is part of Writer's Helper. His program automatically types a series of Xs if the writer takes more than a second between keystrokes.

Blanking the screen to make the text invisible is another funnel device used by WANDAH to force the writer to ignore reviewing and concentrate on planning and translating. The writer cannot review what he cannot read. The aim of invisible writing is to force the writer to put thoughts on paper without worrying about sentence structure, word choice, and other editing concerns. Although invisible writing precludes reviewing, it does not discourage planning, as the flashing screen method does. However, both invisible writing and the flashing-screen method aim to get the writer not to worry about reading and editing the text as he composes, and this is the essence of free writing as the term is used here.

Inventor

To assist with idea generation, a computer should serve as an inventor. Programs that attempt to create, clarify, and order a writer's concepts illustrate the inventor role of idea processors. Inventor devices may be divided into those that aid the writer in forming concepts and those that assist with forming relations among previously established concepts (see Table 16.1). The first approach has been adopted by numerous tutorial programs which were designed to aid students taking college-level rhetoric courses, but could be employed more generally.

Burns (1979, 1984) developed INVENT to serve as a prewriting aid; it asks the writer a series of questions about the subject of the document being composed. INVENT includes three types of heuristics in different programs (TOPOI, BURKE, and TAGI) for different types of writing. TOPOI assists with persuasive writing by using Aristotle's 28 enthymeme topics as the basis for asking questions. The topics are categories of arguments that can be applied to any rhetorical problem (Winterowd, 1968). Corbett (1965) explained that the topics were designed for the composer who has no ideas on a subject, only a few underdeveloped ideas, or a large collection of vague ideas. The topics point to the kinds of arguments that flesh out a thesis. For example, the topics include a concern for the meaning of terms (definition and ambiguous terms), similarities and differences (opposites, correlative terms), reasoning procedures (division, induction), and consequences (simple consequences, crisscross consequences).

BURKE helps with informative writing by drawing on Kenneth Burke's dramatistic pentad. The questions concern scene, act, agent, purpose, and agency (Rueckert, 1963). Burke's pentad is a more sophisticated and powerful version of the journalistic pentad who, what, when, where, and why. The information to be conveyed in a document can be regarded as a dramatic act. To develop the document, it is necessary to specify (1) the scene of the act—where, when, and under what circumstances it was performed, (2) the nature and meaning of the act, (3) the agent who performed the act, (4) the purpose of the agent in performing the act, and (5) the agencies or means by which the act was performed. In addition, it is necessary to explore the relationships, called 'ratios' by Burke, between particular pairs of elements of the pentad, such as scene and act, agency and act, and purpose and act. Burns's (1979, 1984) program is built on ten such ratios.

TAGI aids informative writing by employing the tagmemic matrix of Young, Becker, and Pike (1970). TAGI's questions prompt the writer to see a concept to be included in a document from three perspectives. The concept can be viewed as a particle (as if it were static), as a wave (as if it were dynamic), or as a field (as if it were part of a larger network). These questions follow from only part of the theory proposed by Young, Becker, and Pike. The complete tagmemic matrix consists of nine cells. The rows represent the perspectives of particle, wave, and field. The columns represent three features of any concept to be communicated, namely, its contrastive features (how it differs from other concepts), its range of variation (how it can vary and still remain the same concept), and its distribution in larger contexts (how it fits into a larger system).

To illustrate how these programs operate, consider this brief segment of an actual run of TOPOI (Burns & Culp, 1980, p. 8; student input is in italics):

> During this exploration process, you will be asked to clarify the purpose of your paper on fear of death.
>
> So now would you briefly describe the purpose of your paper by completing this statement: The purpose of this paper is to . . .

Examine the fear of death people have, where we get those fears, and how we may reduce those fears.

Fine, Walt, you and I will talk again about your purpose.

Relax now, Walt, and enjoy this brainstorming session.

What objects do you associate with fear of death?

How might they be included in your theme?

Fear of the unknown, cessation of life, fear of causing pain to our loved ones.

That's the idea, Walt. Give me some more info now.

Pain is another fear causer, knowing we have to die does not relieve the pain/fear of dying.

Super, Walt!

Anything else?

(You can add more info, ask a question, or give a command—whatever you wish.)

How does euthanasia, mercy killing, suicide, murder, and accidental death add to our fears?

I could say that that's for me to know and for you to find out.

Seriously, I cannot pretend to know 'how', but you should keep exploring for an answer.

Continue!

(Try explaining a little more, less phrases, more sentences.)

We're moving right along. Here is Question 2.

What could be considered a result of fear of death?

Nervousness, depression, suicidal tendencies.

By George, Walt, good one. Write a little more, please.

Hatred for the person who caused the death, or the person who died.

Fantastic, Walt! anything else?

Continue!

Okay.

Do you wish to narrow or change your subject?

(Maybe revise the way it sounds in these questions?) (Yes or No?)

Visual synectics is a creative invention heuristic based on drawing analogies between the subject being written about and other seemingly unrelated visual concepts. As with any synectics approach, the aim is to spur creativity by forcing the writer to consider nonobvious relationships (Gordon, 1961). Rodrigues and Rodrigues (1984) described a tutorial program (unnamed) that uses visual synectics. The program first asks the student to name the writing subject. The student then examines a graphic display or a photograph selected from a collection prepared by the instructors. The program asks the writer to describe objects present in the picture. After eliciting a list of five to ten objects, the program asks the writer to compare the subject to each item on the list. The purpose of visual synectics is to elicit novel views of the subject, views the student might never think of without drawing analogies. Rodrigues and Rodrigues (p. 37) gave the illustration of a student writing on nuclear waste disposal who compared the subject to a beachball ('On the surface, the plans seem solid, but there is a lot of hot air underneath that surface') and to a bathing suit ('We may discover that it offers us very little protection').

Related methods for forming concepts are problem statements, nutshelling, and morphological analysis. Young, Becker, and Pike (1970) developed heuristic questions for developing problem statements or for clarifying the subject of the document; these are embodied in Neuwirth's (1984) DRAFT program. Examples of these questions are 'What is the problem?' and 'Are the components of the problem clearly dissonant or incompatible?' Nutshelling is a heuristic developed by Flower (1981) for forming concepts about the rhetorical problem facing the writer. WANDAH (Von Blum & Cohen, 1984) employs nutshelling by asking the writer to state the purpose of and the intended audience for the paper and to provide a synopsis of its main ideas. In short, the writer is asked to 'put it in a nutshell.' Morphological analysis is a heuristic for forming new concepts through a dimensional analysis of old concepts (Stein, 1974). BRAINSTORMER (Bonner, 1984) guides the writer to think of the dimensional structure of two or more concepts concerning the writing subject. The program then establishes a multidimensional matrix of these old concepts. New concepts may be formed through interesting, novel combinations of these dimensions.

A second type of inventor program aims to clarify and order ideas by forming relations among concepts. Smith (1982) described a program (unnamed) that uses a network method to form relations. The program first asks the writer to list the ideas to be included in the text. Then the program presents all possible pairs of ideas, one at a time, and asks the writer if the pair is related. If the answer is yes, the writer is asked to specify the nature of the relation. The program can assist the writer in this by displaying a menu of possible relations (e.g., 'is an explanation of,' 'is analogous to'). After the relations are specified, the program displays in a graphical network each idea as a node, the links among nodes, and a label indicating the type of relation for each link.

A similar program, based on tree structures rather than networks, is called TREE and is part of WRITER'S HELPER (Wresch, 1984). The program asks the writer for a list of ideas and then guides the writer in finding the hierarchical-category relationships among the ideas. After developing the hierarchy, the program displays the resulting tree structure.

Therapist

To deal with affective problems connected with writing, the computer should serve as a therapist. Idea processors that try to reduce the anxiety, frustration, and lack of confidence of the writer serve the therapist function. One way to accomplish this function is to embed therapy within an idea processor whose primary function is to serve as a funnel or an inventor. The therapy delivered in this embedded fashion is covert, in the sense that the writer is not turning to the program primarily for therapy.

To illustrate, INVENT (Burns, 1984; Burns & Culp, 1980) positively reinforces the writer by using terms such as 'good', 'fine', 'terrific', and 'that's the idea' in response to the writer's input. It also makes suggestions to the writer that are primarily affective, not cognitive (e.g. 'Relax now and enjoy this brainstorming session' and 'We'll have a good time thinking about . . .'). Thus, while employing Aristotle's topics as an invention heuristic, the writer covertly receives therapy to alleviate anxiety and build confidence. Positive reinforcement and suggestions of positive affect could be embedded in outlining, free-writing, and other types of idea processors. INVENT seems to be the only extant example of an embedded, covert therapist device.

Alternatively, it is possible to design an overt, independent therapist program that the writer uses with the intention of receiving therapy. Neumann (1986) described several programs that implement specific therapeutic techniques and some of the advantages of such programs over human therapists. For instance, MORTON is a program that delivers Beck's cognitive-behavioral therapy to depressed individuals (Selmi, Klein, Greist, Johnson, & Harris, 1982). Unlike human therapists, MORTON is available any time the user wants it and never gets bored, tired, or angry with the user. As a therapist for writers, MORTON would obviously be most appropriate for those who are depressed. However, because a main tenet of cognitive therapy is the alteration of debilitating and self-defeating thought patterns, it could perhaps be tailored to deal with procrastination, dysphoria, evaluation apprehension, and other symptoms expressed in the self-talk of blocked writers (Boice, 1985).

Although I know of no currently available program, behavioral therapy is another viable method for implementing a therapist device. Reports of success in treating blocked writers with behavioral therapy (Boice, 1982, 1983; Rosenberg & Lah, 1982) indicate that contingency-management software for writers might be worth developing. Such software could be programmed to set up a time schedule for completing the document, monitor the number of words produced per writing session, and deliver verbal reinforcers.

Other points about functions

I regard the funnel, inventor, and therapist functions as the most novel and interesting ways in which idea processors can facilitate planning and translating for writers. But these are obviously not the only such functions. Software can rapidly compute numerical (spread sheets), pictorial (graphics), and linguistic (word processing) information, freeing the writer from the drudgery of doing so. Also, software can store (word processing) and retrieve (text base search) notes, sources, and other information needed by writers. Computation and memory aiding are already well-accepted and appreciated functions. hence, there seems little point in dwelling on them here.

I have discussed the funnel, inventor, and therapist functions as if each program carried out a single function. INVENT illustrates the point that a single program can combine functions (in this case, the functions of inventor and therapist). I assume that the most effective idea processor will be one that integrates all functions. Designing a workable human-machine interface for such an integrated idea processor raises numerous problems that are beyond the scope of the present article. Assuming such an integrated program could be designed, just how effective would it be in helping writers to plan and translate text?

Potential efficacy

Idea processors are still in their infancy. It would be premature to pronounce firm judgments about the efficacy of any of the programs discussed here in helping writers with planning and translating. At this time it is appropriate to address the types of research that are needed, the variables that are likely to be important, and the central theoretical issue that may constrain the effectiveness of idea processors.

Types of research

With one exception, there appear to be no evaluation studies in the literature. The exception is Burns's (1979) dissertation on INVENT. He compared three experimental groups, in which students employed TOPOI, BURKE, and TAGI, with a control group, in which students heard a lecture on the creative process. Burns took several measures of the quality of the students' prewriting inquiry and then of their composition plan (e.g., a detailed outline); the students did not compose a draft for evaluation. He found that all three experimental groups significantly out-performed the control group in terms of the number of ideas generated, the factuality, surprise value, insightfulness, and comprehensive-ness of those ideas, and in the evidence of intellectual processing and overall quality impression of their prewriting inquiry. No significant differences were found on any of these measures among the experimental groups. The quality of composition plans was statistically equivalent for all four groups on all measurements taken. Thus, the benefits seen in the prewriting phase did not carry over to the phase of arranging a plan for a first draft. The attitudes of students toward using the programs were positive. They believed the programs helped them to think and that the heuristics would be useful for many types of writing assignments. More research like Burns's program evaluation needs to be conducted if conclusions about the efficacy of idea processors are to be drawn.

However, considerable conceptual design work is needed before concentrat-ing purely on specific program evaluation, or even program development. The problem is that the best approaches and methods for implementing the funnel, inventor, and therapist functions are unknown. Extant idea processors are based on the conjecture, rather than the fact, that particular methods are powerful controllers of human cognition. For instance, Burns selected Aristotle's topics, Burke's pentad, and Young, Becker, and Pike's tagmemic matrix from a sizeable collection of rhetorical invention heuristics on the basis of Lauer's (1967) theoretical evaluation of the available heuristics. Burns chose the three that theoretically are the most powerful, fully recognizing that empirical validation is necessary. Paper and pencil tests should perhaps be used to determine which invention heuristics merit the expense of program development. Research on approaches and methods of implementing the funnel and therapist functions is also lacking. Such work should specify how to design idea processors for specific types of writers, writing tasks, and writing performance objectives.

Types of writers, tasks and performance measures

A likely outcome of future research on idea processors will be the establishment of boundary conditions on their usefulness. The effectiveness of particular programs will depend on the characteristics of the writer, the writing task, and the performance measure examined. A writer's personality and method of composing may in part determine the usefulness of a particular program. For instance, free-writing programs may have the effect of flustering, rather than freeing, individuals who find it very difficult to ignore reviewing while planning and translating. A perfectionist who insists on trying to compose a polished first draft may be the most in need of free-writing programs and the least able to benefit from them. Similarly, some individuals may prefer the discreetness and anonymity of receiving therapy from a computer, whereas others may regard machines as too impersonal to be of any use. Bridwell, Johnson and Brehe (in press) reported substantial individual differences in the ease with which writers can compose on a word processor. Related differences for idea processors are likely to emerge.

Throughout this article, I have treated writing as if it were a uniform task. Of course, it is not. Scientific, technical, business, journalistic, fictional, and poetic writing, for example, each carry their own unique requirements for being informative, persuasive, and entertaining. Certain inventor devices intuitively seem best suited for particular types of writing, such as Aristotle's topics for persuasive writing and visual synectics for poetry or short stories. In contrast, funnel and therapist devices may work equally well across various writing tasks. Determining the range of applicability of specific idea processors is an important goal for future research.

Lastly, whether an idea processor is effective undoubtedly depends on how effectiveness is measured. Broadly speaking, writing performance can be measured in terms of efficiency and quality. Efficiency refers to the amount of time and effort needed to produce a document. Total time to compose and words composed per minute are examples of efficiency measurements. Quality refers to how effectively a document communicates, and is difficult to quantify (Hirsch & Harrington, 1981). Although no single technique is ideal, the following methods can be employed: holistic judgments of the document (Charney, 1984); judgments of specific features, such as idea development (Atlas, 1979); and calculations of readability based on standard formulas (Klare, 1976). Funnel and therapist devices should have their greatest impact on a writer's efficiency. They seem best suited for reducing the time and effort needed to prepare a document. In contrast, inventor devices should influence the quantity and quality of ideas generated by a writer (see Burns, 1979).

Specific knowledge versus general methods

Although firm conclusions about efficacy are premature, one overarching theoretical controversy surrounding idea processors must be noted. An assumption underlying the use of idea processors is that writers can benefit from the application of general methods of solving problems. Funneling attention to one or two processes, using invention heuristics, and reducing debilitating anxiety are general problem-solving methods that apply across many domains and tasks. If the role of domain- and task-specific knowledge overshadows the role of general methods in writing performance, then idea processors may contribute very little.

Acquiring extensive relevant knowledge about one's language, writing topics, and prospective audiences—and practicing the task of writing for several thousands of hours—may not eliminate writing difficulties, but such expertise should certainly lessen their severity. First, the experienced writer can better handle multiple processes simultaneously, because some aspects of the task are automatically performed. The expert writer may find it possible to concentrate on planning and translating while reviewing automatically, for instance. Dorothy Parker (in Cowley, 1958, p. 10) and William Zinsser (1983), both highly accomplished writers, claimed that they carefully constructed every word, phrase, and sentence as they composed a first draft, enabling them to produce a highly polished piece. Extensive knowledge is undeniably a useful resource for generating good ideas. Hayes (1981) reported that the most creative works of musical composers came only after at least ten hears of intensive study and preparation in the art of composing. This was so even for child prodigies such as Mozart. Lastly, extensive practice at the job of writing—knowing full well the ups and downs to be expected—may help with the problem of emotional hindrance. Having successfully handled procrastination, evaluation anxiety, and other emotional difficulties could prepare the writer for coping with affective interference in the future.

The degree to which general methods add to the contribution of specific knowledge is unclear in the case of writing. The broader issue of how skilled thinking—of which writing is one example—depends on the relative contribution of specific knowledge versus general methods is currently a contested and central issue in cognitive science and engineering (Glaser, 1984; Sternberg, 1985). Designing and evaluating idea processors may be a fruitful way to address this important question.

References

Atlas, M. (1979, December). *Addressing an audience: A study of expert-novice differences in writing.* (Report No. 3). New York: American Institutes for Research.

Beaugrande, R. de (1984). *Text production*. Norwood, NJ: Ablex.

Boice, R. (1982). Increasing the writing productivity of 'blocked' academicians. *Behavior Research & Therapy*, 20, 197–207.

Boice, R. (1983). Clinical and experimental treatments of writing block. *Journal of Consulting & Clinical Psychology*, 51, 183–191.

Boice, R. (1985). Cognitive components of blocking. *Written Communication*, 2, 91–104.

Boice, R. & Johnson, K. (1984). Perception and practice of writing for publication by faculty at a doctoral-granting university. *Research in Higher Education*, 21, 33–43.

Bonner, P. (1984, March). Make a new plan, Stan. *Personal Software*, pp. 120–123.

Bridwell, L., Johnson, P., & Brehe, S. (in press). Composing and computers: Case studies of experienced writers. In Matsuhashi, A. (Ed.), *Writing in real time: Modelling production processes*. London: Longman.

Britton, B. K., Glynn, S. M., Meyer, B. J. F., & Penland, M. J. (1982). Effects of text structure on use of cognitive capacity during reading. *Journal of Educational Psychology*, 74, 51–61.

Britton, B. K., & Tesser, A. (1982). Effects of prior knowledge on use of cognitive capacity in three complex cognitive tasks. *Journal of Verbal Learning & Verbal Behavior*, 21, 421–436.

Burns, H. (1979). Stimulating rhetorical invention in English composition through computer-assisted instruction. *Dissertation Abstracts International*, 40, 3734A. (University Microfilms No. 79–28268)

Burns, H. (1984). Recollections of first-generation computer-assisted prewriting. In Wresch, W. (Ed.), *The computer in composition instruction*. Urbana, IL: National Council of Teachers of English.

Burns, H. L., & Culp, G. H. (1980, August). Stimulating invention in English composition through computer-assisted instruction. *Educational Technology*, pp. 5–10.

Charney, D. (1984, February). The validity of using holistic scoring to evaluate writing: A critical overview. *Research in the Teaching of English*, pp. 65–81.

Corbett, E. P. J. (1965). *Classical rhetoric for the modern student*. New York: Oxford University Press.

Cowley, M. (Ed.) (1958). *Writers at work: the Paris Review interviews* (Vol. 1). New York: Viking.

Daly, J. A. (1978). Writing apprehension and writing competence. *The Journal of Educational Research*, 2, 10–14.

Duncker, K. (1945). On problem solving (L. S. Lees, Trans.). *Psychological Monographs*, 58 (Whole No. 270).

Elbow, P. (1981). *Writing with power*. New York: Oxford University Press.

Flower, L. S. (1981). *Problem-solving strategies for writing*. New York: Harcourt Brace Jovanovich.

Flower, L. S., & Hayes, J. R. (1977). Problem-solving strategies and the writing process. *College English*, 39, 449–461.

Flower, L. S., & Hayes, J. R. (1980a). The cognition of discovery: Defining a rhetorical problem. *College Composition & Communication*, 2, 21–32.

Flower, L. S., & Hayes, J. R. (1980b). The dynamics of composing: Making plans and juggling constraints. In Gregg, L. W. & Steinberg, E. R. (Eds.), *Cognitive processes in writing* (pp. 31–50). Hillsdale, NJ: Erlbaum.

Freedman, S. W. (1983). Student characteristics and essay test writing performance. *Research in the Teaching of College English*, 17, 313–325.

Garfield, E. (1983, April 4). Introducing SCI-MATE—a menu-driven microcomputer software package for online and offline information retrieval. *Current Contents*, pp. 5–15.

Glaser, R. (1984). Education and thinking: The role of knowledge. *American Psychologist*, 39, 93–104.

Glynn, S. M., Britton, B. K., Muth, D., & Dogan, N. (1982). Writing and reviewing persuasive documents: Cognitive demands. *Journal of Educational Psychology*, 74, 557–567.

Gordon, W. J. J. (1961). *Synectics*. New York: Harper & Row.

Gould, J. D. (1980). Experiments on composing letters: Some facts, some myths, and some observations. In Gregg, L. W. & Steinberg, E. R. (Eds.), *Cognitive processes in writing* (pp. 97–127).

Graesser, A. C., Hopkinson, P. L., Lewis, E. W., & Bruflodt, H. A. (1984). The impact of different information sources on idea generation: Writing off the top of our heads. *Written Communication*, 1, 341–364.

Green, D. W., & Wason, P. C. (1982). Notes on the psychology of writing. *Human Relations*, 35, 47–56.

Guilford, J. P. (1967). *The nature of human intelligence*. New York: McGraw-Hill.

Hayes, J. R. (1981). *The complete problem solver*. Philadelphia: Franklin Press.

Hershey, W. R. (1984, May). Thinktank: An outlining and organizing tool. *Byte*, pp. 189–194.

Hirsch, E. D., Jr., & Harrington, D. P. (1981). Measuring the communicative effectiveness of prose. In Frederiksen, C. H. & Dominic, J. F. (Eds.), *Writing: Process, development, and communication* (Vol. 2, pp. 189–208). Hillsdale, NJ: Erlbaum.

Kellogg, R. T. (1983). [Cognitive effort in intentional and incidental learning]. Unpublished raw data.

Kellogg, R. T. (1984, November). *Cognitive strategies in writing*. Paper presented at the meeting of the Psychonomic Society, San Antonio, TX.

Kellogg, R. T. (1985). Computer aids that writers need. *Behavior Research Methods, Instruments, & Computers*, 17, 253–258.

Kellogg, R. T. (1986). *Effects of topic knowledge on the allocation of time and effort to writing processes*. Manuscript submitted for publication.

Klare, G. R. (1976). A second look at the validity of readability formulas. *Journal of Reading Behavior*, 8, 129–152.

Kubie, L. S. (1958). *Neurotic distortion of the creative process*. Lawrence, KS: University of Kansas Press.

Landauer, T. K. (1985). Psychological research as the mother of invention. In Borman, L. & Curtis, B. (Eds.), *Proceedings of CHI '85, Human Factors in Computing Systems* (pp. 1–10). New York: Association for Computing Machinery.

Lauer, J. M. (1967). Invention in contemporary rhetoric: Heuristic procedures. *Dissertation Abstracts International*, 28, 5060A. (University Microfilms No. 68–7656).

Layman, D. (1984, August 7). Framework: An outline for thought. *PC Magazine*, pp. 119–127.

Lowenthal, D., & Wason, P. C. (1977, June 24). Academics and their writing. *London Times Literary Supplement*, p. 282.

Luchins, A. S. (1942). Mechanization in problem solving. *Psychological Monographs*, 54 (Whole No. 248).

Macdonald, N. H. (1983). The UNIX Writer's Workbench software: Rationale and design. *Bell System Technical Journal*, 62, 1891–1908.

Neumann, D. (1986). A psychotherapeutic computer application: Modification of technological competence. *Behavior Research Methods, Instruments, & Computers*, 18, 135–140.

Neuwirth, C. M. (1984). Toward the design of a flexible, computer-based writing environment. In Wresch, W. (Ed.), *The computer in composition instruction*. Urbana, IL: National Council of Teachers of English.

Nold, E. W. (1981). Revising. In Frederiksen, C. H. and Dominic, J. F. (Eds.), *Writing: Process, development, and communication* (Vol. 2, pp. 67–80). Hillsdale, NJ: Erlbaum.

Owens, P. (1984, April). Thinktank and Promptdoc. *Popular Computing*, pp. 186–189.

Rodrigues, D., & Rodrigues, R. J. (1984). Computer-based creative problem solving. In Wresch, W. (Ed.), *The computer in composition instruction*. Urbana, IL: National Council of Teachers of English.

Rose, M. (1980). Rigid rules, inflexible plans, and the stifling of language: A cognitivist's analysis of writer's block. *College Composition & Communication*, 31, 389–401.

Rose, M. (1984). *Writer's block: the cognitive dimension*. Carbondale, IL: Southern Illinois University Press.

Rosenberg, H., & Lah, M. I. (1982). A comprehensive behavioral-cognitive treatment of writer's block. *Behavioral Psychotherapy*, 10, 356–363.

Rueckert, W. H. (1963). *Kenneth Burke and the drama of human relations*. Minneapolis: University of Minnesota Press.

Selmi, P. M., Klein, M. H., Greist, J. H., Johnson, J. H., & Harris, W. G. (1982). An investigation of computer-assisted cognitive-behavior therapy in the treatment of depression. *Behavior Research Methods & Instrumentation*, 14, 181–185.

Seymour, J. (1984, September). Data bases: Managers go on-line. *Today's Office*, pp. 36–40.

Smith, R. N. (1982). Computerized aids to writing. In Frawley, W. (Ed.), *Linguistics and literacy* (pp. 189–208). New York: Plenum.

Stein, M. I. (1974). *Stimulating creativity: Vol. I. Individual procedures*. New York: Academic Press.

Sternberg, R. J. (1985). All's well that ends well, but it's a sad tale that begins at the end: A reply to Glaser. *American Psychologist*, 40, 571–572.

Taylor, C. W., & Barron, F. (Eds.) (1963). *Scientific creativity: Its recognition and development*. New York: Wiley.

Uhlig, R. P., Farber, D. J., & Bair, J. H. (1979). *The office of the future*. Holland: Elsevier North-Holland.

Von Blum, R., & Cohen, M. E. (1984). WANDAH: Writing aid and author's helper. In Wresch, W. (Ed.), *The computer in composition instruction*. Urbana, IL: National Council of Teachers of English.

Walker, L. (1984, April). How to polish up your word processing. *Personal Software*, pp. 108–115, 156–157.

Winterowd, W. R. (1968). *Rhetoric: A synthesis*. New York: Holt, Rinehart & Winston.

Wresch, W. (1984). *Questions, answers, and automated writing. In Wresch, W. (Ed.), The computer in composition instruction*. Urbana, IL: National Council of Teachers of English.

Young, R. E., Becker, A. L., & Pike, K. L. (1970). *Rhetoric: Discovery and change*. New York: Harcourt Brace Jovanovich.

Zinsser, W. (1983). *Writing with a word processor*. New York: Harper & Row.

Suggested further reading is given at the end of Chapter 19.

Chapter 17

Creating Intelligent Environments for Computer Use in Writing*

Lawrence T. Frase

What is an 'intelligent environment'? We know that intelligence has two components—a base of knowledge and a set of procedures for transforming, using, and sharing that knowledge. An intelligent environment, then, is one that has lots of information and ways to change that information and send it around to others. The more knowledge and procedures a system has the more intelligent it is. Furthermore, an intelligent environment is able to coordinate its resources. This conception draws no lines between machines, people, and social systems; each is a resource that contributes to outputs from the system. The question is, how intelligent is educational computing today? To answer that we must look at the computing resources that are now available.

State of the art

This section briefly reviews hardware, software, and network advances. (See Andriole, 1985, for projections of future information processing technology).

Technology

Hardware technology has grown at incredible rates. Since 1970, memory densities and processor speeds have quadrupled about every four years. Output technologies have also grown. For instance, a commercial video disk system allows multiuser access to over 6,500,000 records in the United States Library of Congress shelf list. Compact disks (CD-ROM) are gaining in popularity—about 275,000 text pages fit on one CD-ROM disk weighing less than a third of a pound. Input technologies have grown, too. These include optical character readers, in wide use for the blind, that read aloud the text that is given to them. Developments in optical character reading, speech recognition, and speech production have thus broadened the range of senses of communication via technology.

Progress continues to be made in the development of generalized software tools, such as course authoring systems and software development environments (such as UNIX[1] operating system). These common tool environments are

---

* *Reprinted from* Contemporary Educational Psychology, *1987, 12, 212-221. © Academic Press, Inc. Reprinted with kind permission of the author and Academic Press, Inc.*

important for education, because advances made by one person or project can be shared instantly with others.

Finally, two important technological developments have enhanced human communication. These include advances in network and transmission technology. Local area networks now connect terminals that previously were isolated. Advances in transmission technology have speeded the exchange of information by providing increased rates of transmission, faster modems, and systems that integrate text, speech, and graphics.

In summary, we have computing hardware, software tools, and network resources to alter radically how education proceeds. Now we explore some issues involving the interaction between the emerging intelligent environment and human learner/user.

Theory

Instructional theory. Major processes in learning have been well understood for years. For learning to take place effectively students must (a) produce a response and (b) receive feedback about its adequacy. The difficult pedagogical problem is to manage the complex interactions between student productions and the feedback they get. It is precisely here in the management of complex contingencies that computers can help in creative ways. I give two recent examples because they provide an important contrast in the timing of feedback, which in turn influences the freedom with which a student can operate.

Intelligent tutors provide the first example. Intelligent tutors are based on a theory of errors derived from analysis of student errors or on production rules appropriate to the subject being taught. For instance, the LISP tutor (Anderson, Boyle, & Reiser, 1985) teaches the LISP programming language by checking each segment of code produced by the student for the application of faulty rules. Rule-specific feedback is given when a known error occurs. Here the timing of feedback is contingent on small chunks of behavior (short segments of code). And the subject matter, a programming language, is well defined. But what form of computer interaction might be used to teach ill-defined subject matters such as writing?

This brings us to my second example, *educational advisory systems* (Frase, 1984). Educational advisory systems, as well as intelligent tutors, provide detailed feedback contingent on student response, but the timing of feedback is under student control; hence, a student can compose complex responses before performance is interrupted by system feedback. One example of an advisory system is the UNIX WRITER'S WORKBENCH[2] system (Frase & Dieli, 1986). It includes over 45 programs that identify and comment on specific aspects of a student's writing; however, detailed solutions to writing problems are not given, only advice on possible solutions. For instance, advice to the student might say that difficult sentences should be broken into several sentences, short words substituted for long ones, or passives converted to active sentences. The offending sentences, words, and phrases are highlighted in the student's text, either on paper or on the terminal, and the student has on-line access to advice about correct language usage. Although feedback is available about words, sentences, and paragraphs, the timing of feedback is controlled by the student (in contrast to the timing of feedback in intelligent tutors). For instance, in the LISP tutor students are given feedback after they produce a segment of code. With the WRITER'S WORKBENCH software a student might write several pages before submitting a paper for editorial analysis. With this educational advisory system the activities of writing are not broken into small segments; hence, the computer can deliver detailed feedback, focusing on parts of the text, without disturbing the composition process.

Students in introductory composition who used the WRITER'S WORKBENCH programs for one semester had significantly higher scores on tests of editing than students who did not use the programs (Kiefer & Smith, 1983). Also, they and their teachers had highly favorable attitudes toward the instructional context. Furthermore, a meta-analysis of studies on computer-based instruction shows that adults prefer and profit more from advisory systems than from tutorial systems (Kulick, 1985). Control over the delivery of feedback is apparently important for adults.

Writing theory. Much writing research has been devoted to the process of composition. Cognitive models of the writing process (Flower, Hayes, Carey, Schriver, & Stratman, 1986) specify detection, diagnosis, and revision activities that might profit from computer support. Coupled with work on the structure of texts, representation of knowledge (de Beaugrande, 1980; Kintsch & van Dijk, 1978), and discourse theory (Cooper, 1983; Kinneavy, 1971), cognitive models can be the basis for new instructional procedures (Daiute, 1985; Flower, 1981). For instance, to create an effective automated advisory system one must define the elements of a text that signal writing problems and then clearly describe student actions that can correct them. Cognitive models of writing provide the level of detail needed for computer applications.

The research cited above suggests an outline of the important factors in learning to write. These include getting ideas, organizing thoughts, composing, obtaining feedback about one's writing, communicating with others, and language play activities.

In the section below, I use this outline as a framework for a brief survey of software that may be useful for writing instruction. The section illustrates that we have the computer resources to help with many of the important processes of instruction on writing.

Software

Getting ideas. Expert writers spend about 80% of their time thinking up ideas and organizing thoughts. The act of putting those thoughts on paper hardly reflects the intellectual struggles that have gone before. Prewriting activities, where one invents things to write about, are especially difficult for young children or for those whose knowledge of a topic is weak. The aim of the programs mentioned below is to help writers to start to produce a written product.

Several software programs have been designed to help with the prewriting tasks of planning and generating ideas. Some prewriting programs, for instance the Storymaker program (Bolt, Beranek, & Newman), provide the content, but let the student play with the structure and organization by selecting text elements. One program, called *CAC*, offers children in the lower grades advice about composing persuasive text. The program is based on the assumption that inexperienced children write better if high-level cognitive decisions are prompted by the computer. The child can request computer help by using a special key. This creates a menu of items from which the child may choose. For instance, advice might be sought about choosing the next sentence. The computer suggests actions based on keywords it finds in the preceding text written by the student.

Burns (1984) developed several programs to guide college exploratory writing. INVENT is a suite of programs based on Aristotle's 28 enthymeme topics (TOPOI), Burke's rhetorical pentad (BURKE), and Young, Becker, and Pike's tagmemic matrix (TAGI). The purpose of the programs is to stimulate writing appropriate to different genres. See Burns (1984) for information on how to obtain free copies of INVENT.

Prewriting activities are included in programs such as *Quill*, an extensive program including activities of organizing, composing, and revising text, from D. C. Heath (for grades 3–8), and *Writer's Helper*, from Conduit, which includes 11 different prewriting programs (one, called TREE, displays the tree of ideas developed by the writer) for grades 6 to college.

Finally, about a dozen interactive stories and storymakers are available that might be included under the rubric 'getting ideas'. These include *Storymaker*, from Bolt, Beranek, and Newman (grades 4–8), and *Adventure Writer*, from Codewriter, which allows creation of dialog—style adventures (high school).

Organizing thoughts. Organizing one's thoughts is a second important activity of writing. The computer lets one move text around in various ways and view it from different vantage points; thus it could support organizational activities of writing. Some outline generating programs, such as *Framework* (Ashton-Tate), create empty, numbered outline structures within a word-processing program. Programs like *NLS* expand and collapse outlines. Using *NLS* a writer develops an outline by adding levels to a hierarchical structure with headings such as 1, 1a, 1a1, and so on. These programs have become known as 'idea processors'. Prices for such programs range from less than $100 up to $700. Potential advantages of such programs, for instance, effects on written composition of the ability to view a text from high or low points in its structure, have yet to be tested.

Composing. Technological aids that support the composing process include word-processing programs. There are about 200 word-processing programs that allow children and adults to enter and revise text. Popular word processors, like the *Bank Street Writer* (from Broderbund), present menus of functions from which the author chooses, thus making it easy to learn and to use the system. Many authors, especially adults and professional writers, want to control details of text format, and they need footnote and indexing functions. *Nota Bene* (from Dragonfly Software) is a word processor, but it also permits access to indexed notes and has capabilities for tables of contents, lists, footnotes and endnotes, bibliographies, and indexes. The use of a word processor has obvious advantages for physically revising a text—revisions take about 16 times as long when paper copies must be redone. But there is also evidence that the act of composing is different when students use a word processor. Students write and revise more when they use word processors than they do with paper and pencil (Daiute, 1983).

Aids that support the composing process also include programs whose function is to build poetry or other text. These include *Apple LOGO* (from Apple Computer), which creates random poetry when programmed (grades 4–12), and Steve Marcus's *Compupoem* (University of California, Santa Barbara). Rubin's *Storymaker* program (Bolt, Beranek, & Newman) allows students to create and manipulate text units larger than a sentence. Story structures are represented as a tree consisting of nodes connected by branches. The nodes contain sentences or paragraphs. The student first creates a story by choosing branches to follow; the program adds its text segments to the story as the child chooses. Adults find boilerplate, or preformed text, a useful tool in composing. The point is that computers can do the more routine parts of composition and provide a way to help the user manipulate text more effectively than with paper and pencil.

Feedback about writing. Research has shown that computer programs created to assist the revision process improve student editing skills (Kiefer & Smith, 1983). For instance, the WRITER'S WORKBENCH programs, mentioned earlier, provide feedback on spelling, diction, style, and other text characteristics. An interactive version of the program works within a text editor; it suggests correct spellings for words and will automatically replace them if the author desires. A complete writing laboratory including 12 terminals, printer, and software for managing the laboratory is available from AT&T for less than $30,000. This laboratory is based on trials of the system by the English Department of Colorado State University. Smaller programs, which perform some functions of the WRITER'S WORKBENCH system, are available for less than $100. *The Writer's Assistant*, developed for young children by Levin and others, checks spelling and other features and allows students to try out various sentence combinations.

Communication with others. The importance of social interaction argues against the use of computers as isolated machines. It is through social interaction that human qualities develop, and social interaction is an important component of writing. Feedback from others, by whatever means, develops writing skills. Peer tutoring has been shown to raise the level of writing. Several writing programs, such as the *Quill Mailbag* facility, allow writers to share their products. The *Computer Chronicles News Network* (CCNN) allows children to share news items from around the world.

A system developed at MIT, *Newspeek*, creates personal newspapers. Readers use keywords to describe their interests, and news stories from wire services are filtered according to each reader's list of keywords. Newspapers tailored to each individual are thus produced. An important element in the popularity of computer text processing systems is the ability to format text neatly. With computers, students can publish their own newsletters, papers, or books. For instance, the *MacPublisher* software, with various formats and integration of text and graphics, costs about $150. Publishing their compositions can have strong motivational effects on students.

Language play and motivation. Research shows that students learn most from activities, such as games, that maintain their attention. A host of language games are commercially available. These include games that provide drill and practice for students in enjoyable ways, for instance, *Crossword Magic* from Mindscape creates crossword puzzles from words supplied by the teacher (useful for spelling and vocabulary drills) and *Punctuation Put-on* from Sunburst provides punctuation practice in game format. Tutorial games are available for other language features, such as sentence types and suffixes. Sharples, in the United Kingdom, developed several programs, including *GRAM*, that generate text using rewrite rules. Another program, TRAN, allows students to write their own transformations using pattern action rules. Children apparently learn characteristics of English grammar through such programs; however, we need more research to evaluate the effects of various uses of such programs.

Educational limits of technology

The resources I have reviewed can help writers get ideas, organize thoughts, compose and obtain feedback, communicate with others, and experiment with language. They are a few examples of tools that might be part of an intelligent educational writing environment. But are we using this new technology effectively? I think not. Below, I discuss three reasons why.

Lack of integration

Integration of computer resources is rare in educational computing today. Integration can occur in several ways. One way is to transfer resources among people and places at a particular time. For instance, development of a new word-processing program might be done so that the program can run on several different machines or software developments in one school might be done so that they are consistent and easily transferred (electronically) to other schools in a school district. Such activities concentrate on simultaneous development and access. A second form of integration reaches across time. For instance, hardware designed today should fit with hardware to be designed tomorrow or a word-processing program developed today might be seen as just one part of a larger text managing system for tomorrow.

Third, software can be integrated with other software. But the commercial software that I reviewed is not so integrated as to permit a student easy access to multiple programs at one terminal. Integrated systems require more sophisticated computing resources than the earliest home terminals provide. Research systems, such as the *ZOG* data base system from Carnegie-Mellon University, and the *Hypertext Editing System* and *File Retrieval and Editing System* from Brown University, suggest useful directions for today's small but powerful desktop computers. (See Yankelovich, Meyrowitz, & Van Dam, 1985.)

Finally, setting standards for new technologies is a major issue for education that cuts across national boundaries. Internationalization for an environment, for instance, demands that it be adapted to differences in local customs (dates, abbreviations, decimal delimiters, and so on). International differences in character representation create problems for table lookup and linguistic differences, such as whether a language is read from right to left or from left to right, create problems for text scanning and filtering commands. But without standardization, gains by one educational group are unavailable to others.

Charp (1978) has listed the following qualities of computer-assisted instruction as fundamental concerns of educators: reliability, accessibility, economy, ease of use, compatibility of systems, and available software. One way of achieving these qualities and the types of integration I have just reviewed is to establish shared development environments from which inventions can propagate to others. The most important item for education should be to develop shared environments for use in the schools. A start has been made with the widespread use of the UNIX operating system by educational institutions (Crecine, 1986).

Poor software quality

Hundreds of educational software programs exist, but quality is a problem. Probably less than 30% of commercial educational software in the United States meets minimal standards of acceptability (Komoski, 1985). Although significant advances in industry have occurred in the development of software tools and methods, the fruits of these development tools have yet to be felt in the educational marketplace. More resources should be used to produce high-quality transportable software and to develop shared methods of software testing and evaluation.

Lack of research

A strong program of fundamental research is needed. We need research to clarify the subject-matter domains that we want to teach. For instance, many concepts taught in written composition are difficult to learn because the procedures for using them are not well understood. Improving students' 'voice' in writing is one example, but simpler activities, for instance, correcting overly complex sentences, require explicit statements of cognitive procedures that we do not now have.

Practical questions need answers too. Evaluative research is needed to test the validity and reliability of software. Part of improving software quality would include reliability measures, but programs can work reliably and not be valid. For instance, a program that calculates readability, using an algorithm, might reliably deliver the wrong number (when compared to human scoring) or a program that determines grammatical parts of speech might reliably disagree with human assessment.

Little research has been done on the application of computers to writing. Much of what has been done confounds the causes. For instance, in some studies one cannot tell whether improved writing skills result from the software, the particular way in which the software is used, or from the use of word-processing equipment.

Conclusions and directions for the future

I now return to the question that motivated this chapter. How intelligent is educational computing today? An intelligent environment not only provides instruction for the student but also is an avenue that leads beyond the study hall—a road to new and unpredictable domains of knowledge and to other human beings.

From the technical standpoint, we have a good start on resources for an intelligent environment. Educationally relevant technology today includes much more than computers. A broad range of computing hardware, software, and courseware is available, but we also have new storage devices, such as video disks and compact disk memories, and new means of delivering information, such as cable television and videotex information services. Furthermore, the data-base resources needed to make use of new delivery methods, such as on-line dictionaries, thesauruses, and encyclopedias, are reasonably well developed. In addition, networks allow communication within and across national boundaries. Network and transmission facilities have improved greatly. In short, we have the technology to address the needs of writing teachers, students, administrators, testers, and researchers.

But all these resources have scarcely touched education in America, and they will continue to pass the classroom by until educational environments are structured to coordinate and integrate these resources. In a few universities we see the beginnings of intelligent environments for writing and education in general (Crecine, 1986). We must copy this intelligence from the laboratory into our classrooms.

Notes

1&2. Trademark of AT&T

References

Anderson, J. R., Boyle, C. F., & Reiser, B. J. (1985). Intelligent tutoring systems. *Science*, 228, 456–462.

Andriole, S. J. (Ed.) (1985). *The 'future' of information processing technology*. Princeton, NJ: Petrocelli.

Beaugrande, R. de (1980). *Text, discourse, and process: Toward a multidisciplinary science of text*. Norwood, NJ: Ablex.

Burns, H. (1984). Recollections of first-generation computer-assisted prewriting. In Wresch, W. (Ed.) *The computer in composition instruction*. Urbana, IL: National Council of Teachers of English.

Charp, S. (1978). Futures: Where will computer-assisted instruction (CAI) be in 1990? *Educational Technology*, 18, 62.

Cooper, C. R. (1983). Procedures for describing written texts. In Mosenthal, P., Tamor, L. & Walmsley, S. A. (Eds.), *Research on writing: Principles and methods*. New York: Longman.

Crecine, J. P. (1986). The next generation of personal computers. *Science*, 231, 935–943.

Daiute, C. (1983). The computer as stylus and audience. *College Composition and Communication*, 34, 134–145.

Daiute, C. (1985). *Writing & computers*. Reading, MA: Addison-Wesley.

Flower, L. (1981) *Problem-solving strategies for writing*. New York: Harcourt Brace Jovanovich.

Flower, L., Hayes, J. R., Carey, L., Schriver, K., & Stratman, J. (1986). Detection, diagnosis, and the strategies of revision. *College Composition and Communication*, 37(1), 16–55.

Frase, L. T. (1984). Knowledge, information, and action: Requirements for automated writing instruction. *Journal of Computer-Based Instruction*, 11, 55–59.

Frase, L. T., & Dieli, M. (1986). Reading, writing, and the UNIX WRITER'S WORKBENCH software. *Technological Horizons in Education*, 14, 74–78.

Kiefer, K. E., & Smith, C. R. (1983). Textual analysis with computers: Tests of Bell Laboratories' computer software. *Research in the Teaching of English, 17, 201–214.*

Kinneavy, J. L. (1971). *A theory of discourse*. New York: Norton.

Kintsch, W., & Van Dijk, T. A. (1978). Toward a model of text comprehension and production, *Psychological Review*, 85, 363–934.

Komoski, P. K. (1985, April). *An analysis of elementary mathematics software produced between 1981 and 1984: Has it improved?* A symposium presentation at the annual meeting of the American Educational Research Association, Chicago.

Kulick, J. A. (1985, April). *Consistencies in findings on computer-based education.* A symposium presentation at the annual meeting of the American Educational Research Association, Chicago.

Yankelovich, N., Meyrowitz, N., & Van Dam, A. (1985, October). Reading and writing the electronic book. *Computer*, pp. 15–30.

Suggested further reading is given at the end of Chapter 19.

Chapter 18

Developing a Writer's Assistant*

Mike Sharples, James Goodlet and Lyn Pemberton

An introduction to the Writer's Assistant

The Writer's Assistant is a computer-based *cognitive support system* for people who create complex documents as part of their professional life. It differs from other document creation systems in that it will assist the writer throughout the writing process, from the generation and capture of ideas to the production of a connected piece of prose, combining the facilities of a text editor, an 'ideas processor', and an 'outliner'/'structure editor'.

The system will offer the writer different *views* of the emerging document including:

Notes network view
: This allows the writer to set down ideas as notes and to form these into an associative network.

Linear view
: This shows the flow of text from beginning to end, allowing the writer to perform standard text editing operations.

Structure view
: This allows the writer to create and alter the structure of a linear text, swapping paragraphs, merging sections, and so on.

For each of these views, the writer can choose from a set of *presentations* (such as a *fisheye presentation* (Furnas, 1986) for linear text, where the text at the current focus of attention is shown in full and the text further away is compressed to keywords or section headings, giving both detail and an overview of the document on the screen). The writer will also be able to specify or select *text types* such as 'research paper/in-house style' or 'company report' which will provide a guide to the structure of the document and default constraints such as items to be included, boundaries (such as maximum word length), and context (such as words that should be abbreviated in a consistent manner). Any of these the writer may alter or ignore. As the writer plans and writes, the system will notify her of constraint violations.

Lastly, the Writer's Assistant will allow the writer to annotate her text with comments, markers, notes for revision etc.

---

* *Reprinted from Williams, N. and Holt, P. (Eds.) (1989)* Computers and Writing. *Oxford: Intellect Books, © Intellect Books. Reprinted with kind permission of the authors and Intellect Books..*

Why build a Writer's Assistant?

Writing is a mentally demanding activity. To produce a large document, such as a newspaper feature, a research report, technical documentation, or a company report, the writer will generally need to gather material from a wide range of sources, supplemented by her own ideas and form it into a well-organised text, suited to its audience and purpose. Writing calls on cognitive skills at different levels: information retrieval, knowledge organisation, structural planning, language creation, physical manipulation, perception and comprehension. The main purpose of the Writer's Assistant is to *make writing less demanding* by allowing the writer to collect resources, set constraints, plan, and revise in a single environment.

Although primarily designed to support the writer's familiar strategies, as with any powerful and open-ended software, it will change the nature of the task, as writers discover new routines, short cuts and ways of representing ideas, plans and text. For example, we envisage that writers will construct structure guides and constraint templates for particular types of document, and notes networks for particular tasks, which can be fleshed out as required.

Development methodology

The conventional method of creating a new piece of software is: *specify, implement, test*. The designer produces a specification for the proposed software, describing its functional components and their interrelations, if possible using formal verification methods to prove the correctness of each component, ie. when given the required input it will produce the required results. Programmers then produce an implementation which is fully tested to ensure that the program code accurately meets the specification. This development methodology has been widely advocated for the design of software systems, but it depends crucially on being able to produce a complete and rigorous specification in advance of programming. The methodology is thus suited to software products:

- which implement known algorithms (e.g.. a statistics package)

and

- which support a single, well specified sequence of actions (e.g.. a payroll package)

and

- whose functional components and the operations on them can be specified in advance of implementation.

By contrast, the Writer's Assistant will:

- contain algorithms which are not yet well-specified (e.g.. network management routines and constraint satisfaction routines for text elements)

and

- provide support for sequences of actions (the techniques and strategies of a moderately accomplished writer) which, at the start of the project, were ill-specified and

- offer the writer new ways of working. What the effects of these will be, and whether they will require the functional specification of the system to be modified, can only be discovered once it has been implemented and used in earnest.

For these reasons, the project demands a development methodology which permits a less rigorous initial specification, followed by a process of incremental refinement. Partridge and Wilks (1987) describe the methodology (they call it RUDE for Run, Understand, Debug, Edit) which is commonly used for producing artificial intelligence software. A rough and ready prototype is produced, perhaps with a subset of the necessary functions and a rudimentary user interface, and this is then debugged and refined. Such a methodology is not only suited to the design of AI systems, where there are no standard algorithms and the program itself acts as the specification, but also to 'user friendly' software for a range of different users who cannot articulate all their needs in advance: 'I don't know what I want until I've tried it'.

A RUDE development strategy does not have to be laissez-faire and unstructured. The methodology we describe here (named DCSS, for Development methodology for a Cognitive Support System, pronounced 'decus') is similar to RUDE in that it allows incremental development, but it specifies well-defined design processes and target products. We shall introduce DCSS through the case study of the Writer's Assistant, but it has wider application. Figure 18.1 shows the development of the Writer's Assistant in two main phases. The first

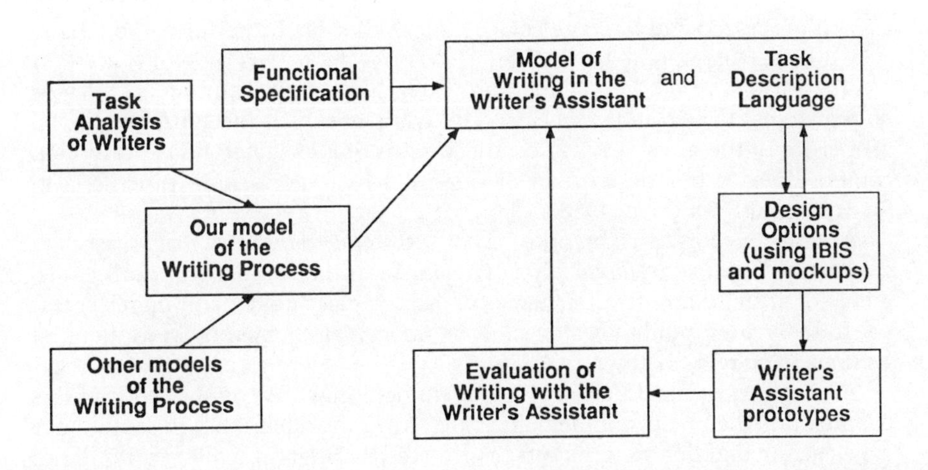

Figure 18.1. The DCSS Development Process for the Writer's Assistant

phase is an analysis of the writing process to produce a *model of writing for the Writer's Assistant* and a *task description language*. The second phase is a specification of software requirements and design options, which are then implemented through a series of prototypes. Tests of the prototypes will lead to the specifications being revised, and to a further cycle of implementation and testing.

The first task is to find a suitable starting point for the process. It is normal in AI software development to either reimplement an existing program (e.g.. the RMYCIN reimplementation of MYCIN (Cendrowska & Bramer, 1984) or to build on an existing model of functionality, generally a model of cognitive functioning produced by knowledge elicitation techniques such as protocol analysis of expert problem solvers (Newell and Simon, 1972), or a task model derived from a task analysis (Johnson, Diaper & Long, 1985).

A model of the writing process

In designing the Writer's Assistant we deliberately sought to break with the tradition of adding further functions and refinements to existing text editors, and to concentrate instead on the writer. We had expected, at the start of the project, that existing descriptions of the writing process (for example from Flower & Hayes, 1980 and Cooper & Matsuhashi, 1983) would serve as the firm foundation for a functional model which could then be implemented and refined. However, these descriptions are of the writer's mental processes and not of the operations, strategies, and techniques carried out on some external medium. This distinction is important because the Writer's Assistant operates not on mental structures but on the explicit external representations that the writer creates. Our concern was to investigate the observable plans and drafts which a writer creates using conventional media, such as pencil and paper or a word processor, and the range of operations that can be performed on them.

We have, therefore, produced an account of the means by which a writer creates a text through interaction with external resources and media such as books, notes, pencil and paper and the ways in which the representational properties of these resources affect the process of idea generation and written composition. A full description of our model of the writing process is in Sharples and Pemberton, 1988.

A writer creates three broad types of text item: *instantiated items* (pieces of connected prose), *uninstantiated items* or *idea-labels* (notes, chapter headings etc. which both stand for mental schemas and act as place holders for pieces of text yet to be written) and *annotational items* (marginal comments, indications of sections to be revised etc.).

An analysis of the different types of annotation is beyond the scope of this chapter. It is the prose and idea-labels that form the material of the text under construction and this material can be arranged in different ways: as unordered collections, as a non-linear organisation such as an 'idea map', or as a linear series. The two dimensions, of instantiation and arrangement, form a *representation space* shown in Figure 18.2 (there is also an arrangement representing the formatted document, but we shall not cover formatting and layout in this chapter). We suggest that what are commonly called 'writing strategies', such as *plan-draft-revise*, and *draft-redraft* and *outlining*, can be described as a sequence of transitions between the different boxes shown in Figure 18.2, where the goal of a normal writing task is to reach box 6. For example, a *plan-draft-revise* strategy is a progression through boxes, 3, 5 and 6, beginning with a non-linear plan which is organised into a linear outline and then fleshed out as a draft.

The great advantage of the 'six box model' is it gives a framework around which to construct the Writer's Assistant. Each box has associated with it a type of representation (idea-labels, notes, notes network etc.), a type of text item (instantiated or uninstantiated), classes of operation which can be performed on the structure of items (generate, verify, transform, instantiate, organise,

Type of Item

Organisation of Items		UNINSTANTIATED	INSTANTIATED
	UNORGANISED	1 *Techniques:* Brainstorming *Representations:* Idea-labels	2 *Techniques:* Note-taking (verbatim) Collecting quotes *Representations:* Notes
	NON-LINEAR ORGANISATION	3 *Techniques:* Following a thread Writing as dialectic *Representations:* Network of idea-labels	4 *Techniques:* Organising notes Filing *Representations:* Network of notes
	LINEAR ORGANISATION	5 *Techniques:* Linear planning *Representations:* List of idea-labels Table of contents	6 *Techniques:* Drafting text Revising text Copying text *Representations:* Linear text

Figure 18.2. The Representation Space for Writing

linearise, shift focus) and constraints on these operations (arising from the task, external resources, language, and context).

Any writing task is governed not only by the representational structures available to the writer, but also by the medium on which they are set down: notes written onto file cards can be easily reordered, text on a word processor can be edited, a sheet of paper or a whiteboard is suited constructing networks of notes, and so on.

The result of our analysis is a model of the writing process with respect to a writer's external symbolic structures and choice of writing medium.

Task analysis of writers

Having developed a preliminary model of the writing process, we then tested and refined it by performing a task analysis of the activities of two people carrying out report writing tasks using both paper and a conventional text editor. A set of core writing actions and objects was derived from the writers' descriptions of task sequences or were inferred from the verbal protocols. The set was then reduced so as to list actions and objects once only, and equivalent items were grouped under categories, with generic terms being chosen to label the categories. The analysis included: knowledge elements required to perform the task which are independent of the writing medium; medium dependent knowledge; the users' goals with respect to tasks (ie. what they say or want can be inferred about the functions needed to achieve a goal); task dynamics (including interruptions, suspensions, digressions, and interleaved tasks). Although a task analysis of two writers could only map out a small part of the space of possible tasks and subtasks it did prove useful information on types of document structure, the writers' resources, and details of the internal structure of notes networks.

Both the writers alternated between low level tasks, such as proof checking, and high level revisions, and the fact that they tended to do the 'easy bits' first implies the need for a flexible writing tool which does not force the writer through items one by one, but lets her take her own path, compiling reminders of items which have been skipped or need further work.

Functional specification

The model of the writing process marks the end of the first phase of the ADROIT development strategy. The bridge to the next phase is a *functional specification* which sets down the essential capabilities of the system, describing the functions and facilities that will be implemented. The conventional tools of writing are quite adequate for a great range of tasks from jotting down a shopping list to writing a novel, and sufficiently familiar that we are not normally aware of their restrictions, but we intend the Writer's Assistant to go beyond passive sufficiency towards active support. It will offer multiple views of the emerging document and provide the facilities of different media, so that the writer is not forced either to select a single medium early in the writing task, or to laboriously copy notes and text from one medium to another.

The functional specification delimits the space of possible designs, without taking positions on particular design issues. ANSI/IEEE standard 830–1984 provides guidelines for a Software Requirements Specification, which includes a functional specification. Although produced for designers of more conventional software (the target product used as an example in the document is a computerised lift system), it offers a useful framework which we departed from only by including an extra section, 'Relationship between Goals and Results', which specifies those aims of the project which depend upon successful research results and may thus be modified in the course of the project.

Model of writing for the Writer's Assistant and task description language

Complementing the functional specification is a *model of writing in the Writer's Assistant* which describes the writer process in relation to the functions provided by Writer's Assistant. Although derived from our model of writing it is more speculative, in that it also describes processes that would normally occur in the writer's head, but which the Writer's Assistant will make explicit, such as the setting and satisfaction of some constraints. (The Writer's Assistant will go beyond other text editing systems in enabling the writer to set down explicit structural, boundary, resource and contextual constraints. The writer might, for example, state that her text should conform to the structure of a research report (with a title and list of authors, followed by optional acknowledgements, an abstract, and a series of sections and ending with a list of references), that it should be a maximum of 2000 words in length, contain quotes from specified sources, and have the phrase 'artificial intelligence' consistently abbreviated to 'AI' and not 'A.I.'). As the prototypes are programmed and tested with writers, then the new model will be revised to account for their range of interactions with the system.

The *task description language* indicates the basic conceptual objects that will be implemented and the range of operations on them.

IBIS and design documents

The functional specification indicates the 'results that must be achieved by the software, not the means of obtaining those results' (ANSI/IEEE, 1984 p.17). It does not, for example, specify how the views and objects will appear on the screen nor how the writer will call up the core operations. Should the writer be presented with different views as overlapping 'windows' or as non-overlapping 'tiles', and if the former, should there be a guard against one view being completely obscured behind another? Should objects be moved by 'direct manipulation' (by being physically dragged around the screen) or by 'indirect manipulation' (with the item and action identified by name)? Should the user choose from a 'menu' of operations, using a mouse or other pointing device, or should there be special keys (marked 'delete', 'copy' etc.) or key sequences to invoke commands? To give informed answers we need to know the space of possible designs, and once decisions are made they need to be recorded, along with their justification, so that if any of them should conflict with some future decision, or cause practical problems in implementation or use (the menus are too long to fit on the screen, or the command key sequences are difficult to remember), then there is rational basis for making modifications to the design.

To explore and record design decisions the project uses a type of Issue Based Information System (IBIS). IBIS was first proposed by Rittel (Rittel & Kunz, 1970) as a method of capturing and structuring argumentation. Rittel proposed that arguments can be represented as arrangements of a small set of *nodes*, indicating *issues* to be debated, standpoints, or *positions*, on the issues, and *arguments*, supporting or challenging the positions. The nodes are connected by *links* (nine types in all) such as *responds-to, supports*, and *is-suggested-by*. Although Rittel conceived of IBIS as a pencil-and-paper technique, there have been recent computer implementations (Conklin & Begeman, 1988) and, in appearance and functionality, IBIS is similar to a 'hypertext' system (Conklin, 1987, provides an excellent survey of hypertext systems).

As a design tool for the Writer's Assistant, we have implemented an IBIS in POP-11 for the Sun workstation. The node and link types have been adapted to the purpose, with four types of node—*Position, Argument, Issue, Comment*—and ten types of link: *Specifies, Generalises, Objects To, Supports, Questions, Contradicts, Responds To, Replaces, Refers To, Comments On*. A particular feature of this implementation is that a number of people can work simultaneously with their own copies of the network: when each person has finished, the system will attempt to merge the different versions.

Each issue in the design of the Writer's assistant is set down as an IBIS issue node (containing 100–300 words of text), to which positions, arguments and comments are then attached. Currently there are some 300 nodes relating to the Writer's Assistant, each one dated so as to give a full history of the design process. For convenience, the IBIS network is periodically 'linearised' and the titles of the nodes printed out to form the basis of paper *design documents* which set out possible options for implementation.

System functionality and mockups

The main guiding principle of the project is that the Writer's Assistant should make writing easier, not harder. Ease of use is an elusive target and, in general, there is no way of predicting *a priori* the effect of a design decision on the usability of a computer system. Building working prototypes is a time-consuming way of testing design options concerned with appearance or simple movement between views. Instead, we created HyperCard™ mockups on an Apple Macintosh™. HyperCard provides a simulated stack of cards, each holding a 'snapshot' of a screen display from the proposed Writer's Assistant, and by selecting 'buttons' drawn on the screen a user can move from one card to another. The effect is like an interactive slide show.

Only the general appearance of the Writer's Assistant can be simulated since the user cannot type in text, nor alter a view, and the screen size and pointing device differ between the Macintosh and the Sun Workstation, but HyperCard mockups were helpful both to answer specific design questions and to get a 'general feel' for how the eventual system might operate. It was particularly useful for designing the views. The three views correspond broadly to boxes 3, 5 and 6 of Figure 18.2, but each has some of the functionality of neighbouring boxes, so as to cut down movements between views. For example, in the notes-network view a writer can set down unorganised idea-labels, prior to incorporating them in the network, and also indicate a linear path through the notes.

Figures 18.3 and 18.4 show examples of the mockup screens, presenting a fictional account of how a typical session with the Writer's Assistant might progress. A writer wants to produce a study of 'Pottery in North Eastern Brazil' so, on entering the system, she selects the option for creating a new document. As she is writing for a newspaper colour supplement she selects *Newspaper article* from a menu of options, getting a structure template and set of constraints appropriate to that type of text. The writer has a fairly firm idea of how the article will fit together, so she starts working in the structure view, sketching an

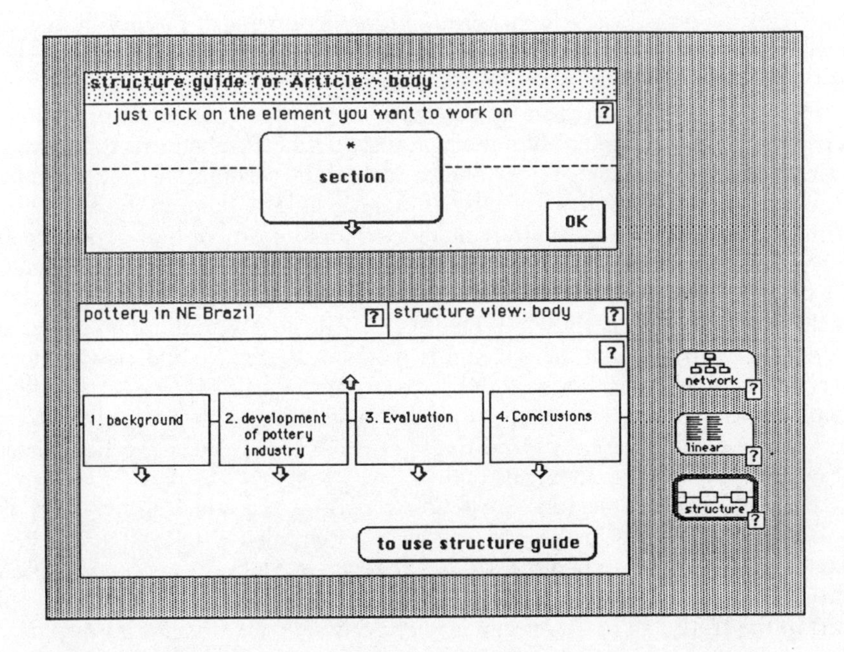

Figure 18.3. Screen Mockup showing the Structure Guide and Linear Structure View

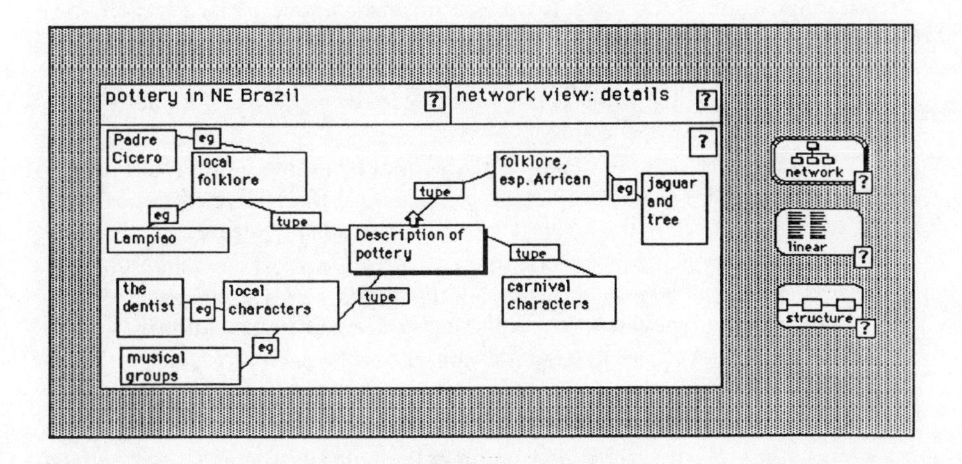

Figure 18.4. Screen Mockup showing the Notes Network View

outline for the body of the text by calling up the *structure guide* to instantiate four main sections headed 'background', 'development of pottery industry', 'evaluation' and 'conclusions'. The relevant part of the structure guide and the resulting state of the structure view are shown in Figure 18.3.

With the overall structure established, our writer chooses to brainstorm ideas for the 'development of pottery industry' section. She turns to the notes network view where she sees a note for each of the sections she created in the structure view. She creates notes for subtopics, linking them to the main note with 'aspect' links. To concentrate on one subtopic—'description of pottery'—she selects a *presentation* of the notes network in which that note is displayed in the centre of the screen and then creates further notes to surround it. The state of the notes network is shown in Figure 18.4. The author might then switch to a linear view to find that the system has made an attempt to linearise the notes network into a list of sections and subsections, based on link types. After making some rearrangements she fills out the section headings with text for the article, moving between different perspectives of the linear view to see sections detail or outline, until the article is finished.

For the experimental study, mockups were designed which showed sample screens at two stages in the production of a demonstration text: at the very outset, and at a point when the text is half drafted. The subjects could explore various views and presentations at each of these two stages. Seven people took part in the study; all of them regularly created complex documents as part of their work and three had extensive experience of using, and in two cases, designing text editing systems. They were first shown a demonstration of the mockups, and then given the opportunity to practice using them for approximately five minutes before being asked to perform a series of tasks such as moving between views and discovering how many sections there are in the body of the text. Finally they were asked a set of general questions and encouraged to offer comments and criticisms.

Without exception the subjects were enthusiastic about the facilities offered by the proposed Writer's Assistant. They understood the rôle of the three views and the structure guide and found the set tasks undemanding. Many of the suggestions made by the subjects were specific to the mockups, but ones which might be more widely applicable include:

- allowing the writer to incorporate material from many different types of document, such as previous reports, source material, electronic mail messages, bibliographic references, other notes and jottings;
- providing a means of maintaining the current context. A writer who is in the midst of a large and complex document needs a 'you are here' indication whenever she moves to a new view or presentation;
- embellishing the 'scroll bars' (these indicate the current position in a long document) with a graphic representation of the document's structure so that the writer can scroll more easily to a desired position;
- providing a 'cut ends' presentation of the notes network, allowing the writer to focus on a defined set of notes;
- allowing the writer to suspend the current activity so as to experiment within a view;

- providing 'shortcuts' for expert users of the systems, such as keypress sequences which parallel the mouse and menu commands;
- providing help messages in a predetermined section of the screen simultaneously with the current document, rather than in a separate window which obscures it.

All the subjects noted that while they are pleased to find that the system is designed to support their current writing practices, the provision of a novel type of writing tool such as the Writer's Assistant will necessarily encourage new writing strategies. The implication is that it will be just as important to study and support the techniques of writers composing with prototype versions of the Writer's Assistant as to support conventional strategies.

Cycle of implementation and evaluation

With a clear specification of the purpose, functionality, and user interface for the Writer's Assistant, the goals of the implementation are well-defined, but achieving them is by no means simple, since the system will have novel features for which there are no standard algorithms. These include:

- automatic transformations between views. The most difficult will be to devise a method of generating a linear view from a notes network, finding a route through the notes which produces an acceptable first draft;
- detecting conflicts between the emerging text and the constraints which the writer previously set;
- determining a convenient time to interrupt the writer. Too many interruptions will result in the writer losing the thread of concentration; too few and the text may have moved so far from the constraints that revision is difficult;
- providing ready access to other documents created using the Writer's Assistant and merging documents with differing structure and constraints;
- maintaining the writer's context when moving between views.

As Figure 18.1 shows, there will be a cycle of implementation, evaluation (by means of a task analysis of writers using the successive prototypes) and revision to the design. The evaluation may also suggest the need to revise the model of writing in the Writer's Assistant as writers discover new techniques and possibilities for writing. At the time of writing this chapter, we are at the stage of implementing the first prototype.

Conclusions

The main purpose of this chapter has been to suggest that the development of computer-based tools for writers need not be a choice between, on the one hand, adding further features to existing text editors and on the other hand, an *ad hoc* approach to system design based on intuition, trial and error. We have set out a structured development methodology (DCSS) which is consistent with the techniques of AI and the design of interactive systems. It is currently being put to the test in the development of the Writer's Assistant system. Writing is only one of many complex, demanding tasks that can be made easier with computer support, and a goal for future research will be to apply the methodology to the development of other types of cognitive support system.

Acknowledgements

Financial support for the Writer's Assistant project is provided by a grant from British Telecom. Claire O'Malley worked as a consultant to the project and carried out the task analysis.

References

ANSI/IEEE (1984). *ANSI/IEEE Std. 830–1984: IEEE Guide to Software Requirements Specifications*, July.

Cendrowska, J. & Bramer, M. (1984). Inside an expert system: A rational reconstruction of the MYCIN consultation system. In O'Shea, T. & Eisenstadt, M. (Eds.) *Artificial Intelligence: Tools, Techniques, and Applications*. New York: Harper and Row.

Conklin, J. (1987). Hypertext: An introduction and survey. *IEEE*, September.

Conklin, J. & Begeman, M. L. (1988). gIBIS: A hypertext tool for exploratory policy discussion. *MCC Technical Report Number STP-082-88*. Software Technology Program, MCC, Austin, Texas.

Cooper, C. R. & Matsuhashi, A. (1983). A Theory of Writing Process. In Martlew, M. (Ed.) *The Psychology of Written Language*. Chichester: John Wiley.

Flower, L. S. & Hayes, J. R. (1980). The Dynamics of Composing: Making Plans and Juggling Constraints. In Gregg, L. W. & Steinberg, E. R. (Eds.) *Cognitive Processes in Writing*. Hillsdale, NJ: Lawrence Erlbaum.

Furnas, G. W. (1986). Generalised Fisheye Views. In *Proceedings of CHI'86*, 16–23.

Johnson, P., Diaper, D., & Long, J. (1985). Task Analysis in Interactive Systems Design and Evaluation. In Mancini, G., Johannsen, G. & Martensson, L. (Eds.) *Analysis, Design and Evaluation of Man-Machine Systems*. (Proceedings of the 2nd IFAC/IFIP/IEA Conference).

Newell, A. & Simon, H. A. (1972). *Human Problem Solving*. Englewood Cliffs, NJ: Prentice-Hall.

O'Malley, C. & Sharples, M. (1986). Tools for Management and Support of Multiple Constraints in a Writer's Assistant. In *People and Computers: Designing for Usability*. (Proceedings of the Second Conference of the British Computer Society Human Computer Interaction Specialist Group). Cambridge: Cambridge University Press.

Partridge, D. & Wilks, Y. (1987). Does AI have a Methodology which is Different from Software Engineering? *Artificial Intelligence Review*, 1, II, 111–120.

Rittel, H. & Kunz. W. (1970). Issues as Elements of Information Systems. *Working Paper #131*. Institut für Grundlagen der Planung I. A., University of Stuttgart.

Sharples, M. & O'Malley, C. (1988). A Framework for the Design of a Writer's Assistant. In Self, J. (Ed.) *Artificial Intelligence and Human Learning: Intelligent Computer-Aided Instruction*. London: Chapman and Hall.

Sharples, M. & Pemberton, L. (1988). Representing Writing: an Account of the Writing Process with Regard to the Writer's External Representations. *Cognitive Studies Research Paper No. 119*. School of Cognitive and Computing Sciences, University of Sussex.

Suggested further reading is given at the end of Chapter 19.

Chapter 19

WordProf
A Writing Environment on Computer*

Maria Ferraris, Francesco Caviglia, and Riccardo Degl'Innocenti

Introduction

'A major breakthrough in the teaching of writing has been made possible by the convergence of two recent developments in science and technology . . .' This is the beginning of an insightful 1980 paper which introduced the idea of a computer-based system bringing together text handling facilities and theoretical knowledge on the cognitive processes involved in writing. The envisaged system was a 'Writing Land' capable of training students on specific aspects of the writing activity, as well as assisting them in the actual writing process (Collins and Gentner, 1980).

Since then the idea of computer support for the teaching of writing has gained in audience and credibility. In the last few years a number of experiments have been carried out, mostly on the use of word processors (WPs) for text composition; the results, however difficult to quantify, all suggest that the computer may play a positive role in enhancing communication skills and reflecting on one's own writing strategies. The first meeting between computer and writing is likely to develop into an intriguing marriage, for several reasons.

The two advantages pointed out by Collins and Gentner are far more real now than they were in 1980: on the one hand, today's WPs provide the writer with increasingly flexible, powerful and easy to use text handling capabilities; on the other hand, much research work in the field of cognitive sciences and artificial intelligence has promoted a deeper and better formalized knowledge of the cognitive processes involved in communication.

If we look at its diffusion, the computer is today an all pervasive instrument, due to its manifold use as a medium in information technology (consider, for example, E-mail, desktop publishing, or group coordination systems). Its pervasiveness has two major consequences. First, good communication competences are gaining a renewed importance at social level. Second, the use of computers for writing, reading and communicating is likely to induce into the cognitive processes and the products of these activities some changes, the characteristics and significance of which are not clear yet. However, these

* *Reproduced from* Educational Training and Technology International, *1990, 27, 1, 33-42 with kind permission of the authors and the Association of Educational Training and Technology. Educational Training and Technology is published by Kogan Page Ltd. on behalf of AETT, BMA House, Tavistock Square, London WC1H 9JP, England.*

changes do exist, and authors such as Garcia Marquez, Primo Levi and Umberto Eco report how their style and production have been affected by the use of the WP.

As for writing itself, two more suggestions from studies in language and cognition contribute to give particular significance to the teaching of writing. First, not only is writing a communication tool, but also it is a means of organizing knowledge and improving the quality of thinking. Second, writing is a problem-solving activity that requires very complex cognitive processes and strategies to tackle at the same time content, style and language constraints (see, for example Gregg and Steinberg, 1980; Scardamalia and Bereiter, 1985 and 1986).

Teachers, for their part, are recognizing communicative competences as a key to those 'learning to learn' skills whose acquisition is a major objective of the educational system in the information society, and whose lack is judged as a main cause of school failure. Hence the growing attention paid by the school to fostering reading and writing skills (in the case of Italy this is significant: only in the last few years is the teaching of composition and comprehension being taken into account, albeit seldom practised, in the secondary level curriculum).

In summary, we are facing two trends: (1) the computer endorses new writing technologies and determines a new need for communicative skills, and (b) cognitive and instruction sciences point towards the development of strategies and tools in order to support the teaching of writing and to reduce the cognitive load of the learner.

But how can a 'writing land' for 'apprentice writers' be developed as suggested by Collins and Gentner? Attempts in this direction have already been made and are discussed in this chapter. WordProf, courseware based on a WP with integrated educational functions and a set of activities devised for a student writer, is then focused on. Before describing WordProf and discussing its theoretical assumptions, we will briefly explore today's main trends in the use of computers in the teaching of composition.

Trends in teaching text production with the computer

Studies and experiences in this field tend to overcome the traditional computer-assisted instruction (CAI) approach (the computer as a patient and monotonous trainer in low-level language skills) and refer instead to the use of office automation tools, e.g. WPs, or to the development of software environments where students are stimulated to carry out communicative experiences or to reflect on their own writing strategies.

Research and field test appear to follow three main guidelines:

- the computer as a *tool* (ie the use of commercial WPs in teaching how to write at various school levels);
- the computer as an *expert* (which implies software with language and text competences);
- the computer as a *teaching environment* (which implies a software with educational features to assist the student in the various phases of producing text).

These trends are not neatly separated, but it is useful to consider them one by one.

The computer as a tool (ie the word processor)

WPs offer numerous educational possibilities, for example:

- the editing tools foster a view of the text as a plastic material to be easily shaped through successive refinements, where reviewing and revising result seamlessly integrated in the writing process;
- the writer's detachment from the typed text (as opposed to the hand-written one) should promote a more objective assessment of the work, as the writer alternates in the roles of the author and the critical reader;
- the text written on the screen and then printed out on paper is intrinsically public domain, thus promoting an experience of co-operative writing (writing *with*) and a better perception of the communicative function of a written text (writing *for*).

But are these real advantages? In other words, does something really change when a WP is used to teach how to write? How can we fully exploit the possibilities offered by this tool? Which WP is needed, with what features? These are the main issues faced by this approach.

Several researches point towards some common results after the medium-term use of WPs: 'low-level' errors (mainly spelling) decrease; the text is more often revised (but there is no agreement on what kind of revision is actually performed); the length of the written text increases; most students show a positive attitude towards writing. (For an analysis of the results, see Potter, 1987 and Calvani, 1989.)

On the other hand, it is rather difficult to evaluate and explain qualitative differences of the produced texts, as well as to measure modifications in students' awareness about their production strategies. Even though the obtained results are not always so easy to decode, it is commonly agreed that the use of the WP often increases the quality of writing. It is more difficult to ascertain to what extent such improvements are due to the WP itself or to other factors. As Potter suggests, the introduction of a new tool tends to modify the teaching approach. In the case of teaching to write, the WP implicitly leads to a re-drafting approach and therefore '. . . any observable improvement in the children's writing could be due to the re-drafting approach rather than the word processor' (Potter 1987).

Other studies leave out the question of measuring the possible advantages of the WP and simply take these advantages for granted: in this approach the focus is rather on how the WP should be used. In this case a reviewing and rethinking process of the writing curriculum in face of WPs may lead to new activities (co-operative writing organized across local and remote networks, text production for real-world purposes, composition based on pre-existent texts) or to a different organization of old ones.

Another point of view focuses on defining the features of a WP for use in an educational context. In some cases, these studies have led to the development of new WPs designed specifically for school use—a major example is WordWise Plus.[1] In our experience with students at different school levels, the most significant features have proved to be exactly the same as for good commercial software: ease of use and power—ie no awkward combinations of keys to perform standard editing, WYSIWYG, speed, text-graphic integration capabilities (see Degl'Innocenti and Ferraris, 1988).

The computer as an expert

The attempts to provide a computer with language and text competences have brought about software tools that are able to operate, in a more or less intelligent way, on language data: some of these tools (spelling-checker, on-line thesaurus, hyphenation and so on) are already available in many word processors; others still belong to AI research about natural language 'understanding' (for a list of the available software, see Lancashire and MacCarty, 1988). Of particular interest for this approach are some systems meant to act as critical revisors. For example, Writers WorkBench (Cherry, 1980) and Epistle (Heidorn et al., 1982) are able to spot some syntax errors and style inadequacies (mainly for business letters, in Epistle), and to suggest some solution through a system of dictionaries, morphosyntactic parsers and rules.

Another text revisor worth attention is the last version of Grammatik III, which has the advantage of running on personal computers.[2] Grammatik III, a product that embodies some AI techniques, submits texts written on different WPs to a critical analysis, searching for low-level errors (capital letters, repetitions, repeated punctuation), style errors (on the basis of a dictionary of overworked, trite or hackneyed phrases), syntax errors (subject-verb disagreement, incomplete sentences, improper use of articles). Grammatik III claims to be 'the easiest way to improve your writing', pointing out that, even though not designed for school purposes, the student writer could be one its possible users.

It is questionable, however, if the error-based approach alone may help a student to develop composition skills that in this case are more a pre-requisite. Text analysis tools could rather gain in educational significance where integrated in an environment (not necessarily computerized) designed to assist the whole process of text production.

The computer as a teaching environment

The idea of assisting a student during the various text production phrases has led to the development of specific-purpose educational software.

The new release of Writer's Helper, software first published in 1980 and recently updated, could be seen as a first translation (albeit limited) of the already mentioned idea of a 'Writing Land': Writer's Helper assists students, through various strategies, in the pre- and post-writing phases, providing them with suggestions to explore and organize ideas, as well as tools for revising the text (word frequency, detection of inadequate forms and terms, etc.).[3]

Writer's Helper assumes, in the software implementation, a clear cut between pre- and post-writing and does not envisage any intervention during the actual writing, while a more recent trend points toward computer environments capable of dynamically assisting the student writer *during* the composition; in fact, some new tools in WPs such as 'outline' or 'revision marks', can be very suitable for this purpose.

The proposal for a 'writing assistant' advanced by Sharples and O'Malley (1988) gives a major contribution for this approach. It outlines the features of a software environment which

> ... should offer to the writer three separate but mutually consistent views of the emerging text, in the form of an ideas structure, a stream of text and a layout. The writer should be able to move between them at will and, in altering one view, know that the others will remain in step. Give one view, the system should be able to generate approximations to the other two.

Beside the different views on the text, the proposed system should automatically take on low-level tasks (formatting, manipulating word and sentence syntax etc.) in order to reduce the writer's cognitive load. The purpose mentioned by the authors is to help the student to develop control and reflective skills, ie the meta-cognitive skills to cope with the writing task with regard to goal, content and language constraints.

Bryson and Scardamalia's 'Muse' likewise aims at developing meta-cognitive skills, but its instructional strategy implies a more explicit teaching role. Muse is a writing environment 'designed to teach students strategies for sustaining independent reflective inquiry during the composition of argument type compositions'. Muse does not teach the student the structure of a good argumentation, but rather stimulates the use of heuristic strategies by explicitly showing the procedures an expert writer follows in structuring argumentations: through activities such as Challenge Assumptions, Identify Confusion, Planning, New Ideas etc., Muse aims at inducing awareness and developing skills in these strategies. Muse is numbered by its authors among those 'computer supported intentional learning environments [that] foster rather than presuppose the ability of students to exert intentional learning control over their own learning' (see Bryson and Scardamalia, 1988).

A computerized writing environment: WordProf

WordProf, a joint project between CNR and IBM Italy, can be located across two of the above described approaches: the WP as a tool and the computer as a writing environment. The aim of WordProf is to devise courseware capable of supporting and stimulating the students who are learning to write *while* they are writing. A single environment should therefore incorporate both the functions of a standard WP and some tools and functions dedicated to the 'apprentice writer'. But what are the tools and functions that could be added to a WP in order to turn it into an *educational environment?*

To answer this question we have started analysing written text from different viewpoints: as a product, as a process and as a set of educational problems. The result has been a network of the activities involved in written composition, including those allowing a control over the whole process.

The network, a fragment of which is shown in Figure 19.1, has been used as a general reference framework for developing courseware that will eventually cover all nodes of the network by means both of tools for performing certain activities and of advice or training on those nodes regarded as the most critical for the apprentice writer. This process has led to add to the standard WP functions, the specific menus illustrated in Figure 19.2.

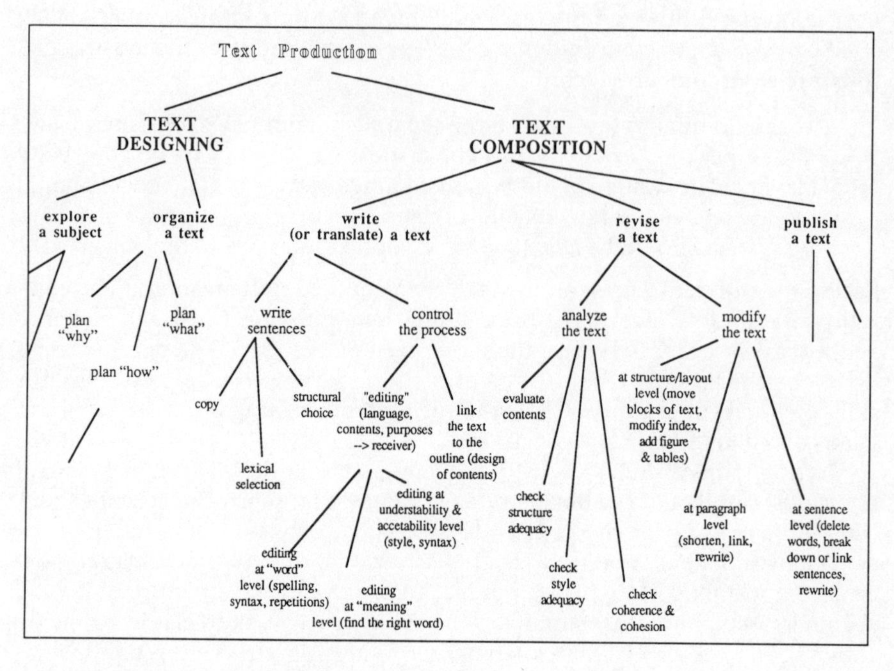

Figure 19.1. A fragment of the hierarchy

The student using WordProf works with a word processor which, as well as the standard File, Edit, Find, Format functions, includes additional menus specifically designed for educational purposes. The whole could be regarded as a writing 'atelier' and the added components as special 'rooms' of this atelier: a *lab* (organized into tools and view menus) furnished with a kit of statistics tools and text-viewing/revising facilities; a *library*, where the student can access a database of pre-written texts, a style dictionary and a general handbook; the *assistant's* room, that provides (to a limited extent) ideas and suggestions for the various writing phases; and finally a *gym* to train some basic skills.

The software also includes a simple authoring system allowing the teacher and the students to enrich and modify both the set of gym exercises and part of the contents associated with the items listed in Figure 19.1. One teacher guide and one student guide will come with the software, conveying both theoretical analysis and practical suggestions about composition strategies. The whole WordProf package is therefore structured as in Figure 19.3.

The 'education rooms' of WordProf will be examined briefly (for more details on the various items, see Caviglia, Degl'Innocenti and Ferraris, 1989).

The lab contains *tools* and *workbenches* (Tools and View menu, respectively) to act on the text as a physical entity. The 'tools' include: *markers*, to underline particular words (connectives, gerunds etc.), spot repetitions, highlight beginning/end of each paragraph, and *counters*, to measure the sentence and paragraph length and to build a frequency index for all words in the text or for predefined (possibly user-defined) word sets. The *workbenches* (View Menu) are devised to allow different views and working modes on the text as a whole: through write comments a revisor can attach to a critical spot notes and

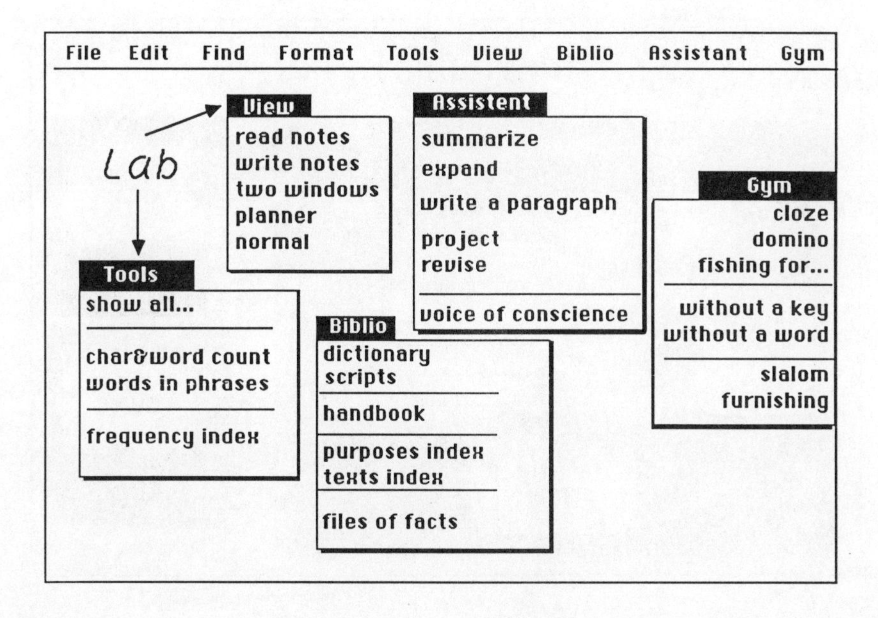

Figure 19.2. The menus of WordProf

suggestions without modifying the original text; *read comments* is useful for viewing someone else's comments while modifying a copy of the original text; *two windows*, with a quick-transfer facility, is devised mainly for note-taking; finally, *the planner* is an outline facility for working on the text at different levels of detail.

The *library* contains a database of texts that students can consult as information sources or copy into the WP for their own purposes. Students can find: *reference works* such as a dictionary of synonyms and a grammar/style handbook with access both through a table of contents and through a list of 'problems'; *scripts* made of words and phrases organized for semantic fields; a supply of *texts* which can be accessed by kind (letter, report etc) or by style features or even according to different *communicative goals* (in the documentation, some activities are strongly suggested that lead students to adapt these models for different needs and situations); *files of facts* on different topics, in order to foster the habit to a documented writing.

The third room belongs to a not too brilliant, but helpful, *assistant* whose task is to provide, on request, ideas and help on various activities that can be carried out during the writing process: summarizing or expanding a text; sketching an outline; writing individual paragraphs; revising a paragraph, an outline or a whole text. In some cases the assistant starts a dialogue with the students in order to put forward, on the basis of their answers, suggestions on the most adequate text form (e.g. letter, report), style and discourse structure, or it can provide dynamic examples of how a certain activity can be performed, or it can even suggest some guided activities or refer the student to the Tools or Library menus, if they are deemed useful to the solution of the students'

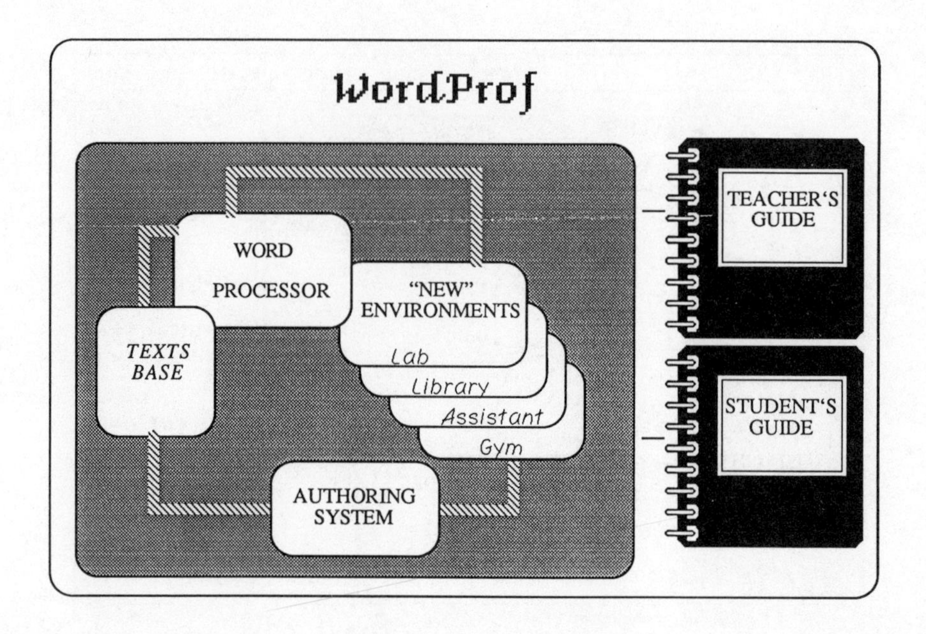

Figure 19.3. The main components of WordProf

problem. The assistant also includes a revisor which, if activated, can interrupt the writing on its own authority to point out some kind of inadequacies (over-lengthy paragraphs, lack of punctuation, overuse of some linguistic forms, repetitions): hence its name 'voice of conscience'.

In the *gym* the student can train (in some cases freely, in others under the control of a coach) with games aimed at developing basic writing skills: syntax and vocabulary adaptability (*cloze, slalom, without a key, take out word, furnishing*) and text comprehension (*fishing* means taking notes, *domino* refers to text or outline re-ordering). Some of the games deliberately break some writing conventions that students adhere to, forcing them to develop new ones: for example, *take out word* eliminates invariable speech parts from the students' text (suppose, the connective 'and') and prevents them from using such words on rearranging the text; *slalom* 'shoots' randomly chosen connectives into the WP text, forcing the students to use them while writing on a given subject.

Even if the features added to the standard WP seem to perform very different functions, many of them are linked to each other by means of a two-fold connection, internal and external to the software. Internally, some menu items automatically use others (e.g. by touching any word of the frequency list all occurrences of that word in the text can be retrieved, or else the assistant can provide 'check-up' of the text by automatically using the functions of the Tools menu). As for the external connections, the various items, functions and working modes have all been derived from a global analysis of the 'text production' process (the network in Figure 19.1) and only afterwards have been organized according to functional affinities: that network is based on the instructional activities, requiring a combined use of several tools and features, proposed in the accompanying booklets.

WordProf: background assumptions

Any tool, let alone a software package, assumes a more or less explicit model of the work it is devised to perform. For example, a word processor can be said to embody a dynamic logic of written composition, while educational software such as *Writer's Helper* seems to assume a stage-model of writing. What, in this respect, are the assumptions underlying WordProf?

The project is based on considerations and choices on writing models, on instructional approaches to composition, on the role to be assigned to the computer. A first choice links any instructional support to writing with the act of writing itself, seen as the complex set of activities leading from an idea or a communicative need to a text. A second remark concerns the dynamics of the writing process: numerous studies show how the various writing stages do not follow a linear sequence but occur with interruptions, iterations or recursions within a highly dynamic process (see, for example, Hayes and Flower, 1980). It is therefore possible and maybe even useful to train students to operate in stages (pre-writing, writing, post-writing), but the educational writing environment where they work must not bind them to this single strategy.

The WP is an ideal tool to put this dynamism into practice, since it integrates all text production functions into a single tool. Through it the writer starts designing the text, outlines it, writes and refines it and edits it, always in the same working environment. However, the WP embodies no educational competence; learning to write, on its side is so complex an activity that it would be naive to envisage it as a learning-by-discovery process, and it would be at least unfair not to provide students explicitly with those contents (rules, heuristic strategies) which can help them to cope with the cognitive difficulties involved in writing.

These remarks underlie one of the basic choices of WordProf: a writing environment which is primarily a WP, integrated with a set of educational functions acting inside the WP and not at its borders, so that students can move freely, while writing, between the typical tools of a WP and the added educational ones.

Another fundamental choice has been to keep the whole courseware flexible, in order to make it usable for different learning paths, chosen either by the teachers in accordance with their goals, or by the students in accordance with their communicative needs and personal writing strategies, just as the same WP suits different texts and writers. These two choices require the integration of the instructional components into the computer, without fixing one exclusive strategy of use, because there is actually more than just one writing strategy. This is not a trivial goal and has probably been attained only partially in WordProf. Whereas the *tools*, the *views* or the texts in the *library* assume an instructional theory but do not directly privilege one strategy rather than another, for the *assistant* (and for some games under control) the instructional design is unavoidably explicit and can prove conditioning. To minimize this effect the most directive *rooms* of the systems are 'open', that is, they can be expanded and modified by the user. Furthermore, the explicit teaching of different writing strategies is reported mainly in the teacher and student booklets associated with the software.

As for the teaching approach, we generally try to encourage: writing for 'real' purposes (real-world situations and text models); multi-hand writing/revising processes; imitation of 'good' text models, rather than laying stress on detecting errors and inadequacies (though spot-the-error exercises can easily be implemented with the authoring system).

WordProf: project development state and problems

The WordProf prototype, which is at present being developed, runs on IBM/MS-DOS systems as a Microsoft Windows application written in the author language of 'The Whitewater Group'. The completion of the evaluation with secondary school and university students is expected by the end of 1990; both the software and the written material will be successively refined and expanded.

Up to the present, different problems have been faced in implementing the project. As for hardware and software choices, we must stick with the currently available/affordable technology, since our purpose is not the product in itself but its use; at the same time we must also face the obsolescence risk if we fail to foresee correctly the evolution of computers in the next few years. Hence, for example, the decision has been made not to bind this software to any current commercial WP, but to program a very simple one and to develop the additional features in such a way that in the future they could be integrated into more powerful WPs.

Another set of problems concerns educational issues and results from the difficulty of clearly expressing and systematizing (not to mention transferring it to a computer) a kind of knowledge about strategies and processes which are not yet sufficiently known and which contain many elements hard to formalize. To make things even more difficult, it is not easy to envisage now how the regular use of computers will actually modify cognitive and practical aspects of writing. These last educational issues lead us to some concluding remarks on the role of the computer in teaching to write.

Conclusions: Which role for the computer?

The contact points between computer and writing teaching that we listed at the beginning of this chapter can all be encountered in the stimulating idea of co-operative writing between a human author and an artificial mind, as suggested by the Italian psycholinguist Parisi a few years ago. Parisi devised an 'expert in the screen' capable of intervening, after and during the writing process, to correct errors or to detect inadequacies as well as to suggest modifications, but capable also, from a language and textual knowledge base, of interacting with the author 'already in the initial phases, even before the text itself, or a part of it, is laid down' (Parisi, 1987).

This is an intriguing perspective which, taken to its extremes, might even eliminate the need to learn how to write texts. Such an expert might allow for automatic text production starting from some contextual indications (purposes, contents, target etc.). This perspective is quite remote and therefore hardly useful for educational purposes, in view of the difficulties (a) of being theoretically unbeatable, according to some, and (b) of developing systems able to dialogue in natural language (actually in this case much more would be at stake, since the competences required to co-operate in producing a text go well beyond mastering the language).

The concept of 'co-operative writing', however, is the best description of the possible role of the computer in teaching to write. It embodies the ideas of 'environment', 'tool' and 'expert', and to some extent does away with the need to distinguish between them; at the same time, it requires understanding the terms of this co-operation, ie defining how each part contributes to the whole process. At the moment, as regards writing and learning to write, we think that the computer can perform very well some useful and 'non-intelligent' tasks (ones that do not actually affect either text semantics or teaching strategies and do not require *real* competences). This involves some consequences. First, the computer should be used in a learning context capable of providing those portions of linguistic and instructional 'intelligence' which the computer does not have. Therefore, while developing and using educational software, as well as word processors, electronic dictionaries or text revisors, much attention should be paid to devising global learning paths in which the computer is inserted *as well*. Second, not only should new software reproduce on the computer strategies and tools born independently from electronic writing, but the peculiarities of the computer as a tool must also be taken into account, and an effort made to devise how possible changes in the writing process could affect learning and teaching to write.

Notes

1. Megarry, J. and Deeson, E. (1988) *Wordwise Plus Literacy Pack*. Special Needs Software Ltd., 74 Victoria Crescent Road, Glasgow, G12 9JN.
2. Wampler, B. E. and Williams, M. P. (1988). *Grammatik III*. Reference Software Inc., 330 Townsend Street, Suite 123, San Francisco, CA.
3. Wresch, W. (1988). *Writer's Helper: Stage II, CONDUIT*. University of Iowa, Oakdale Campiu, Iowa City.

References

Bryson, M. and Scardamalia, M. (1988). MUSE, a computer based learning environment for novice and super-novice student writers. In *Proceedings of ITS-88*, Montreal.

Calvani, A. (1989). Didattica della scrittura con il word processor: aspetti teorici e applicativi. In Calvani, A., (Ed.) *Scuola, Computer e Linguaggio*, Turin: Loescher.

Caviglia, F., Degl'Innocenti, R. and Ferraris, M. (1989). WordProf: an atelier for the apprentice writer. In *Proceedings of 'Reading, Writing and Word Processing'*, (in press). Ormskirk: Edge Hill College.

Cherry, L. L. (1980) *Writing Tools: The STYLE and DICTION Programs*. Bell Laboratories Computing Science Technical Report No. 9. Murray Hill, NJ: Bell Laboratories.

Collins, A. and Gentner, D. (1980). A framework for a cognitive theory of writing. In Gregg, L. W. and Steinberg, E. R. (Eds.), *Cognitive Processes in Writing* pp. 51–72. Hillsdale, NJ: Erlbaum.

Degl'Innocenti, R. and Ferraris, M. (1988). *Il computer nell'ora di italiano*. Bologna: Zanichelli.

Gregg, L. W. and Steinberg, E. R. (Eds.) (1980). *Cognitive Processes in Writing*. Hillsdale, NJ: Erlbaum.

Hayes, J. R. and Flower, L. S. (1980). Identifying the organization of writing processes. In Gregg, L. W. & Steinberg, E. R. (Eds.), *Cognitive Processes in Writing*. Hillsdale, NJ: Erlbaum.

Heidorn, G. E., Jensen, K., Miller, L. A., Byrd, R. J. and Chodorow, M. S. (1982). The epistle text critiquing system. *IBM System Journal*, 21, 305–326.

Lancashire, I. and MacCarty, W. (1988). *The Humanities Computing Yearbook*. Oxford: Clarendon Press.

Parisi, D. (1987). Verso un sistema esperto per la elaborazione di testi. In Poggi, I. (Ed.) *Le parole nella testa* pp. 401–410. Bologna: Il Mulino.

Potter, F. (1987). *Word Processing and Literacy Skills: An Outline of our Current State of Knowledge*. Commission of European Communities, GN2438-86-11-NIT-UK.

Scardamalia, M. and Bereiter, C. (1985). Development of dialectical processes in composition. In Olson, D. R., Torrance, N. and Hildyard, A., *Literacy, Language and Learning*. Cambridge: Cambridge University Press.

Scardamalia, M. and Bereiter, C. (1986). Research on written composition. In Wittrock, M. C. (Ed.) *Handbook on Research on Teaching*. New York: McMillan.

Sharples, M. and O'Malley, C. (1988). A framework for the design of a writer's assistant. In Self, J. *Artificial Intelligence and Human Learning*. London: Chapman and Hall.

Suggested further reading for this section

Britton, B. K. & Glynn, S. M. (Eds.) (1989) *Computer Writing Environments: Theory, Research and Design*. Hillsdale, N.J.: Erlbaum.

Clark, R. E. (1985). Confounding in educational computing research. *Journal of Educational Computing Research*, *1*, 2, 137–148.

Kellogg, R. T. & Mueller, S. (1990). A knowledge-based view of composing on a word-processor. Paper from the authors, Dept. Psychology, University of Missouri-Rolla, Rolla, MO 65401, USA.

Nickerson, R. S. & Zodhiates, P. P. (Eds.) (1988) *Technology in Education: Looking Toward 2020*. Hillsdale, N.J: Erlbaum.

Sharples, M. & Pemberton, L. (1990). Starting from the writer: guidelines for the design of user centred document processors. *Cognitive Science Research Report No. 154*, School of Cognitive & Computing Sciences, The University of Sussex, Brighton, BN1 9QN, U.K.

Williams, N. (Ed.) (1990) *Computer Assisted Composition*. Oxford: Intellect Books.

Williams, N. & Holt, P. (Eds.) (1989). *Computers and Writing*. Oxford: Intellect Books.

List of contributors' addresses

Part 1

Chapters 1 & 2	James Hartley, Department of Psychology, University of Keele, Staffordshire, UK, ST5 5BG

Part 2

Chapters 3 & 4	Collette Daiute, Graduate School of Education, Larsen Hall, Harvard University, Cambridge, MA 02138, USA
Chapter 5	Susan M. Zvacek, Department of Educational Technology, University of Northern Colorado Greeley, CO 80639, USA
Chapter 6	Stephen A. Bernhardt, Department of English, New Mexico State University, Box 3E, Las Cruces, NM 88003, USA

Part 3

Chapter 7	Charles A. MacArthur, Department of Special Education, University of Maryland, College Park MD 20742, USA
Chapter 8	Keith E. Nelson, Department of Psychology, Pennsylvania State University, 417 Bruce C. Moore Building, University Park, PA 16802, USA
Chapter 9	Richard Ely, Division of Special Education & Rehabilitation, School of Education, Boston College, Chestnut Hill, MA 02167, USA
Chapter 10	Alan J. Koenig, College of Education, Texas Technical University, Box 4560, Lubbock, TX 79409, USA
Chapter 11	Sydney J Butler, Department of Language Education, University of British Columbia, Vancouver, BC, Canada, V6T 1Z5

Part 4

Chapter 12	Gary R. Morrison, Department of Curriculum and Instruction, Education Building 424, Memphis State University, Memphis, TN 38152, USA
Chapter 13	David Kember, Educational Technology Unit, Honk Kong Polytechnic, Hung Hom, Kowloon, Hong Kong

Chapter 14	Ben Shneiderman, Human-Computer Interaction Laboratory, Department of Computer Science, University of Maryland, College Park, MD 20742, USA
Chapter 15	Charles B. Kreitzberg, Cognetics Corporation, 55 Princeton-Highstown Road, Princeton Junction, NJ 08550, USA

Part 5

Chapter 16	Ronald T. Kellogg, Department of Psychology, University of Missouri-Rolla, Rolla, MO 65401-0249 USA
Chapter 17	Lawrence T. Frase, Educational Testing Service, Princeton, NJ, USA
Chapter 18	Mike Sharples, School of Cognitive and Computing Sciences, University of Sussex, Brighton, UK, BN1 9QN
Chapter 19	Maria Ferraris, Istituto per le Technologie Didattiche, CNR, Via Opera Pia 11, 16149, Genova, Italy

Subject Index

Author Index